Snafu Snatchers

Recovery of Downed Airmen

13th Air Force
2nd Air Sea Rescue Squadron

Clark Field, Philippines
1946

Grey T. Larison

authorHOUSE

AuthorHouse™
1663 Liberty Drive, Suite 200
Bloomington, IN 47403
www.authorhouse.com
Phone: 1-800-839-8640

© *2009 Grey T. Larison. All rights reserved.*

No part of this book may be reproduced, stored in a retrieval system, or transmitted by any means without the written permission of the author.

First published by AuthorHouse 1/29/2009

ISBN: 978-1-4389-4493-7 (sc)

Printed in the United States of America
Bloomington, Indiana

This book is printed on acid-free paper.

Produced by

Nature Episodes Inc
1218 Auburn Rd
Box 57
Locke, New York 13092

Front cover: PBY or Catalina
Back cover: Tribute to Rescue Personnel

Dedicated to my fellow airmen who performed
Life saving Air Sea Rescues in the South Pacific

Clark Field, Philippine Islands 1946

Their devotion to duty and willingness to perform difficult, dangerous rescues saved the lives of countless airmen and other travelers when other transportation failed them.

Writing this book required gathering material from various sources including narratives of others with WWII veterans and official military records especially mission reports.

PBY Catalina or Cat

B 17

Foreword

My heart goes out to everyone that has endured the trials and tribulations of war. And I mean everyone. As described in the upcoming pages of "Snafu Snatchers," the realities of war seeps throughout each thread of the fabric of our existence, and still does. It affects a clean clear pool of water like drops of crimson dye slowly blending in to change forever. You'll feel the beginnings of this as you read from Grey's account of his life growing up in Odessa, N.Y. in WW II Begins, and continues throughout the book as he recounts himself and other younger men that "grew up" quickly while struggling through compelling experiences as medics and rescue workers during the war in chapters Manila, Philippine Islands and the 2nd Air Sea Rescue Squadron.

Although much has changed in the realm of emergency services today compared to care that was given during Grey's account during the Snafu Snatchers era—or even just two years ago for that matter, one thing is timeless and all encompassing; we are all just doing the best we can.

Within my heart lives an oxymoron, my gratefulness to the men and women of the armed forces who fought and continue to fight for this country, and the game that exists through the politics of war itself. Anonymously said, "War is a matter of old men's pride and young men's lives. You, the reader are fortunate to read this book as it is penned with the voice of a gentleman that I can visualize sitting right next to me. Knowing Grey and his wife, Marie for over a decade now, knowing their passionate love of life, read on and find out for yourself the devotion to humankind, the intimate, disarming and daunting.

Elizabeth Heath
Ithaca, New York

Contents

Foreword		ix
Glossary of Terms		xiii
World War II Begins		1
Endurance and Survival		3
Military Mobilization		16
My Service Begins		34
Training for Service		38
A Sea Voyage		51
Manila, the Philippine Islands		59
2ND Air Sea Rescue Squadron		87
The Squadron Area		103
Favors Negotiated		108
Mess Hall Operations		110
Snafu's		122
March 17, 1946	A Long Taxi Ride Across Open Water	127
March 20, 1946	Surviving G.I. POW's	130
April 2, 1945	A Leader of Men	135
April 10, 1946	The Mountains on Mindoro	138
April 21, 1946	A Jungle Trek	143
May 2, 1946	An Engine Change	155
May 11, 1946	A Call for Help from a Foreign Ship	158
May 21, 1946	33 Days on a Raft	164

June 5, 1946	Snafu One of Our Own	173
June 11, 1946	Extracting Wounded Soldiers	176
June 11, 1946	Extracting Fighting Soldiers	179
June 21, 1946	A Dangerous Snafu on Angry Seas	184
June 30, 1946	A Murdered Filipino	191
July 6, 1946	The True Measure of a Commander	192
July 12, 1946	Exploring a Remote Beach	195
July 19, 1946	A Pregnant Patient	199
July 24, 1946	Jungle Rot Plagues Snafu Crews	203
August 6, 1946	Late Arriving Air Crews	205
August 16, 1946	Typhoon	206
August 22, 1946	A New Base Commander	210
August 31, 1946	Not All Snafu's Have Happy Endings	211
September 8, 1946	An Accident on Clark Field	214
September 21, 1946	Take Training Seriously	215
September 27, 1946	Shopping in Manila	217
October 14, 1946	A Moonlight Flight	220
October 21, 1946	Death in the Moonlight	226
November 1, 1946	Jap Soldiers Continue Fighting	231
November 22, 1946	Refugee Horses	239
December 10, 1946	"Jungle Jim" Says Goodbye	244
The Return Home		246
Epilogue		257

Glossary of Terms

Aircraft mentioned in Snafu Snatchers
    AT6 Advanced training aircraft.
    B17 The workhorse bomber Four engine capable of lifting thousands of pounds of cargo (bombs).
    B29 The largest bomber used during WW11. Carried the atom bomb to Japan.
    Catalina, Cat or PBY A twin engine amphibian aircraft used as a patrol and for search and rescue.
    C45 A twin engine cargo airship usually carrying passengers only.
    C46 A twin engine, larger size cargo airship.
    C47 A twin engine medium size cargo airship.
    C54 Skymaster A four engine passenger-cargo airship used for long distance hauling.

Atabrine: a drug developed to replace quinine to better control. Malaria. A pill, usually three times weekly.

Bivouac: A temporary camp, with or without tents.

Blister: A bubble of transparent Plexiglas, often raised from the fuselage to form a bubble or blister.

Buck a line (chow line). Attempt to cut in front of soldiers already waiting in line for something like food, drinks, mail or service of any kind.

Carbine: Light weight, short .30 caliber semi-automatic combat rifle. Generally used in closeup fighting.

Direction terminology used to describe a direction from a ship or aircraft. When facing in the direction of travel.
>Aft: Behind or astern
>Amidships: The center of a ship.
>Bow: Straight ahead.
>Fantail: The overhanging deck on the aft of a ship.
>Port: To the left.
>Starboard: To the right.

Dutchman: a life boat attached to a B17 rescue aircraft can be dropped to survivors on the water.

Feather a Propeller. Turn propeller blades so they do not bite into the wind, they may or may not turn without pulling or retarding the aircraft.

Fantail: The main or after deck on a ship.

Forty or eight (40 or 8) a type of railroad car developed during WWI, means 40 men or 8 horses

Grease gun: a 45 caliber machine gun, very light weight easily handled, a throw away weapon

Gibson Girl: A radio with a 200 foot wire antenna attached to a kite to enable a radio signal to reach distant installations.

Gook: A term applied to all Asians. This was not necessarily intended to be a derogatory or negative term.

Grease gun: a fully automatic machine gun using 45 caliber ammunition. Resembled a grease gun used to lubricate automobiles at the time.

Garand 30 caliber semi-automatic rifle widely used by Infantry troops during WWII

Huks: Communist Filipinos supplied by Communist China attempting to hijack the newly formed Democratic government of The Philippines.

Mae West: An inflatable life vest worn by airmen. The inflatable chambers filled in front of the wearer's chest, hence the name.

MOS number: A number assigned to military personnel after training designating skills acquired.

Mariana Islands: A group of islands 3,000 miles east of The Philippines.

Music Paper: Toilet paper

Negritos: Short black person of the Austronesian region.

Non-Com's Non Commissioned Officers.

Jap: This term was applied to all Japanese soldiers, who were hated. When spoken, soldiers would use it in a harsh manner, spitting the word out.

Panzers: German tanks, used in North Africa, Russia and Europe

Rations:
    K Rations. Bricks of packaged condensed food easily carried and or rehydrate. Can be eaten either hot or cold. Each box contained sufficient food for one meal, about 1500 calories. Packed in breakfast, lunch or dinner menus.
    C Rations. Cans of food. This could be most anything. ie. meat, vegetables or fruit.

Ruptured duck: Emblem sewn over the pocket, on the right side of the Ike jacket signifying the wearer has an Honorable Discharge from the Army.

Sex Education:
> Mickey Mouse Films: Sex education Army style. Monthly Short Arm Inspection: Visual inspection of genitals seeking the presence of venereal disease.

Very Pistol: Fired a flare into the sky,

Whores bath. Also called a spit bath. Wash from water contained in a small bucked like a helmet.

Wehrmacht: The Regular German Army.

World War II Begins

I first learned about the coming war from the back seat of my father's car. Mom, Dad, my younger brother and I first read Pearl Harbor had been bombed on the evening of Dec 7, 1941. We had been visiting my grandmother, Ethel Goit Williams LaValley, in the St. Joseph Hospital in Elmira, New York at the time. "Monnie" was slowly dying from bone cancer.

My grandmother Ethel LaValley died January 9, 1942. We buried her the January 12, 1942 in Woodlawn Cemetery, Elmira, New York. She was never told of the attack on Pearl Harbor.

Leaving St. Joe's Hospital, Dad drove past The Star Gazette building on Baldwin Street Sunday December 7, 1941 to check the headline bulletin board placed in a window so it could be read by passing motorists. Posted on the large bulletin board were exceptionally large and bold headlines announcing the attack on Pearl Harbor. It was late in the afternoon as we drove along the streets in Elmira. Newsboys were appearing on the street, hawking an "extra edition".

The next day we listened to President Roosevelt's radio address broadcast from the House Chambers on Capital Hill, when he uttered those never to be forgotten and emotional words, "This day will live in infamy" and a "state of war" exists between The United States and Japan speech on 8 Dec. 1941. That very day mobilization began. Congress immediately declared war on Japan. Hitler declared war on the United States a day or two later.

The entire world immediately erupted in a bloody brutal war, lasting more than four years, with huge costs in both lives and treasure for all involved.

Grey T. Larison

Dad (Ted Larison,) never served in the Military. He was too young to serve during WW I. He participated in other ways. As a teen aged Boy Scout he raised a Victory Garden on his parent's property each year, his father owned about 5 acres on 17th street in Elmira Heights, New York. Actually his Victory Garden was the project for his Eagle Scout Badge.

Dad was a very enthusiastic Boy Scout and extremely proud of his Eagle Rank. His was the fourth awarded in the Sullivan Trail Council recently formed in Elmira.

During WW I food was in short supply. America was soon to become the bread basket for the Allies. Home owners who had available land were encouraged make Victory Gardens to grow food. These foods were to be canned or dried for seasonal use. Calculations indicate Victory Gardens during WW II, accounted for 40% of the civilian food supply.

Endurance and Survival

My father owned and operated a grocery store on Main Street in Odessa, New York, one he purchased in January 1938. Dad had previously been a traveling salesman the previous 18 years for The Salada Tea Co. Dad called on small grocery stores throughout the Southern Tier of New York, checking them for stock, making displays and distributing sales materials.

During these years, Dad developed a strong desire to own his own store. Opportunity came to him when an independent store in Odessa became available. There were competitive stores on Main Street, Odessa, including a small regional chain outlet. Dad soon made his the biggest store, having the most in sales and serving the greatest number of people in the Odessa shopping area.

With the event of Pearl Harbor, the entire country was immediately united in the cause of destroying the yellow bastards. Deep hate for anyone of Japanese ancestry instantly developed. All citizens began to make sacrifices, food, gasoline, tires, clothing and building materials were severely rationed. People were making deep sacrifices dedicated to support the war effort. Causality reports began to arrive, almost daily. Severely wounded men were returning home, some to die, others to recuperate, some remained disabled for the remainder of their lives.

On the home front production of all material goods immediately began on a grand scale never previously achieved. There was an immediate demand, for large quantities, of airplanes, tanks, trucks, rifles, ammunition, clothing, food and ships or any other item used to prosecute a war. Every person, man, woman or youngster was urgently needed in some capacity or another.

This was the beginning of great social change that only became recognized until much later long after the war ended. Women began leaving their homes or small farms seeking work. Some began searching for their own professions thus changing their entire family relationships. This social change was to continue for the next several decades as women, sparked by this opportunity, sought equality with men in the workplace.

Early in the war wheels were set for complete social change that was to become apparent at wars end. Women had become breadwinners for a family with their own money. Having acquired this power and authority women were not willing to give it up when husbands returned. Returning soldiers found family conditions very different than when they left.

The full mobilization of the country soon produced shortages in nearly all consumer goods. Automobiles no longer came off the assembly lines in Michigan. The last models produced for sale to the public were in December 1941, vehicles already in existence be kept serviceable for the duration of the war.

Coal, widely used in homes at the time, was trucked to Odessa from coal mines in northeastern Pennsylvania. Supplies were uncertain as there was such heavy demand for it. Fuel, mostly coal, was allocated to businesses with government contracts on the basis of need for the war effort with little remaining for home heating. Heating oil, for homes was also rationed and very expensive. Pipelines moving gasoline, fuel oil or kerosene from Texas and Oklahoma had not been constructed, making it necessary to truck all oil products by tanker truck to Northern States. Coal and oil was both strictly rationed and not available.

Rubber tires were immediately no longer available to the public. Rubber tree sap (latex) was collected from trees grown in tropical areas, mostly Brazil and Southeast Asia.

Latex, taken from rubber trees was shipped to the United States where it was processed into rubber used for making tires. With the outbreak of hostilities, raw latex shipped from Southeast Asia stopped immediately when the Japs invaded Malay and Indonesia.

The remaining source for raw latex was from Brazil, the only area where latex was available to the United States. It was needed in large quantities or an immediate alternative must be found.

Fortunately Henry Ford had invested considerable capitol to develop rubber tree plantations in Brazil. Increased demands created by wars do stimulate inventions and development of new products will cause profound changes.

It had been known for some time how to make synthetic rubber, but it had never been brought to market. With the war and it's extremely high demand for vehicle tires both civilian and military, synthetics were developed and went into mass production. Synthetic rubber has been developed and improved over the years since war's end to become much superior to natural rubber.

The Ford plant at Willow Run Michigan was converted from producing autos to producing heavy bombers, especially the B17. Farm equipment such as tractors, plows, rakes and wagons were no longer available for the civilian market.

While the skyrocketing demand for food kept increasing, all food produced on American farms had to be grown and harvested with existing equipment.

This temporarily ended the move towards mechanizing small farms. Horses again came into general agricultural use and extensive use for short haul transportation. There were two horse traders in the Odessa area that made regular trips to the Midwest to purchase both draft and saddle horses. For several war years they continued to operate thriving businesses,

Aluminum was soon unavailable for civilian use as most current production was sent to aircraft factories. Pots and pans already in retail stores disappeared quickly.

One might ask what about plastics? Prior to Pearl Harbor, few plastics had been developed or sold on the general market. Cotton and wool was diverted to make uniforms.

While recycling was not the buzz word at the time that is just what Americans were doing as a matter of survival. The words used were "wear it out, make it do, or find another use for materials". For example; Mrs. Rosekrans, who had a family of several children, took jeans that were worn down the front, the knees worn through, pockets torn, might still have some use. Mrs. Rosekrans noted the backs of the legs were in comparatively good condition. She took the jeans apart, salvaged the back leg material to reuse in sturdy quilts. Thus Mrs. Rosekrans and others coped with the lack of new fabric and bedding.

Cloth sacks coming to farms containing grain were saved, disassembled, dyed and made into various items of clothing. It was not unheard of, when a couple were to be married, her wedding dress might be made of grain sacks, cut, dyed and completely refashioned for the occasion. That same material might be reused to make other garments.

Lawns were plowed; gardens planted in their place. The Victory Garden concept of WW I was resumed, however the term was rarely used. Even in the village of Odessa, in addition to backyard gardens, chickens were raised in small coops beside most homes.

While chickens needed some whole grain, they could also forage off the land eating whatever lawn and weed seeds they could find, resulting in very little or no expense to raise. From these free roaming chickens came both meat and eggs for human consumption.

Many neighbors so inclined and possessing the necessary space and skills, kept a milk cow or two, thus supplying the family and perhaps some neighbors with fresh milk and butter.

A standard workday was 9 hours or 45 hours per week. Many came home after a full days work and spend another three or four hours caring for animals and gardens.

People did what they had to do to prosecute the war. Americans were united in the purpose of defeating both Japan and Germany as they have never been united since. During the war years a dissenting word was scarcely uttered.

Gallons of blood were collected and made available to hospitals and the Red Cross. Blood was fractionalized into components. Plasma was sent to battle zones and given as soon as possible to wounded soldiers. The needs were acute. It became apparent from the very first battles, wounded soldiers sometimes by the hundreds, and were in need of that life giving plasma.

Immediately, rationing became well organized. Gasoline was the first to become strictly rationed. The Government ruled each registered, licensed automobile would be issued one ration book for gasoline containing one coupon to be used each week. The value of a coupon could be changed by the Government each month. Initially a coupon was good for four gallons of gasoline. Automobiles at the time averaged about 14 miles per gallon.

That was not enough gasoline for a farm family to make one trip a week to town for groceries, visit a Doctor, family or a visit to the Post Office.

Car pooling became necessary and expected. If a person needed more gasoline, they applied to the Ration Board for additional allotments. For example a person working in a defense plant in Elmira needed to commute 40 miles or so each day was allowed additional gasoline but additional allotments granted were never sufficient to cover necessary driving.

The only answer was car pooling, regardless of how inconvenient. Neighbors might theoretically be able to ride together to Elmira where most war production and jobs existed. However, neighbors might work on changing swing shifts or in different factories. Coordinating time and distance to save gasoline was difficult, if not impossible in most situations. Factory workers were mostly married women with children. Baby sitters must be scheduled before mothers could start work outside the home. These were extremely difficult situations to work out in a practical manner.

Immediately food became strictly rationed. Each man woman and child was issued a ration book. Ration books were issued, one each, to every man, woman and child.

Inside were coupons for fresh and canned goods, coffee, sugar, lard, cheese, eggs, red meat and poultry. Each ration book contained the same number of stamps. An infant had the same ration as an adult, far more than a young child could eat. This meant a family with 4-6 children had plenty of food.

The government calculated a person would be able to receive 1700 hundred calories daily from ration books. This is not enough nourishment for a physically active, working adult.

Most men, under 38 years of age were eligible for the draft into Military Service. At war's beginning pre Pearl Harbor fathers were exempt from Selective Service but that was soon to change.

The few men with agricultural deferments worked farms producing agricultural commodities. Agricultural deferments were granted only to those working on high production farms. In the Odessa area these were farms producing dairy, fruit poultry and eggs or other livestock. Marginal farm operations did not qualify men for draft deferments.

Fresh produce from local farms or gardens sold directly to consumers were not rationed. Most eggs produced on a local farm or in someone's backyard came from small flocks of less than a hundred birds. Many families, even those living inside the village of Odessa, kept chickens. Surplus eggs were traded for other foods in local stores so were almost always available. Spent hens and cockerels provided an additional source of food..

Those few larger farms producing poultry specialized in the broiler market, not eggs. Broilers could be brought to market size, about 2 1/2 lbs, within 12-14 weeks, consuming less grain than laying hens.

Locally produced fresh milk and butter was also available. Like eggs, milk was produced in someone's backyard. It was customary for many village residents to keep a cow or two. From these cows came fresh milk, butter, sometimes cheese. Excess calves, generally males, would be fed extra to promote rapid growth, later they were butchered to supply a family with enough meat for nearly a year.

Excess home produced food was traded at local grocery stores for such items as coffee, sugar, chocolate, spices or salt, items impossible for small family operations to produce. Milk, butter and eggs were all rationed.

Retail food stores were required to collect Ration Stamps and could neither pay nor collect more than the ceiling prices for any of these items.

Government bureaucrats fool themselves if they think something like this will happen within the law as they have written it. During wars or natural disasters, people will do whatever is necessary to survive.

This was the beginnings of the so called 'black market'. We soon learned 'the so called black market' may exist for survival, not necessarily with profit making the only motive.

People having no land area where they could produce some of their own food were immediately helpless. This included the many families working for Cotton Hanlon a local firm that cut logs into lumber.

Most Cotton-Hanlon laborers lived in Cayuta, New York, where sawmills were located. They occupied company housing with no land area available for either gardening or livestock.

Then there were also elderly and disabled, people living on the very meager dole the County Welfare gave them. As I observed many of the older generation, with no family remaining, probably had multiple disabilities and illnesses were warehoused in some falling down shack of a house, not fit for human habitation. This was done with the full knowledge of the County Welfare System, with or without a war.

It was almost impossible for people so treated to get enough food. Poor nutrition almost certainly results in further physical deterioration, hastening both the onset of further disease and /or death.

Dad was concerned exactly how some families would be able to survive. It had always been his priority to physically go to people's homes to take their orders for groceries. Groceries were delivered to their home the next day. This took gasoline. With the coming of rationing the nagging question rose?

Where to find gasoline enough to serve these disadvantaged people? It seemed a way must be found somehow to "Work the System" if all were to survive. How to get food produced on local farms to those who had the greatest need. Ration Boards issued gasoline stamps for the production of food.

OK they make the rules, let's make them work for us. Encourage disabled or elderly farmers, not working the soil or keeping livestock to apply for gasoline rations. It was easier for farmers to be issued gasoline ration stamps providing the farmer had a few acres and a tractor or other farm equipment on his farm. Dad would either buy or trade for those gasoline ration stamps and use them in getting food and other supplies to customers.

Ration Board regulations called for gasoline intended for use on farms must be delivered to a storage tank located on the farm land where it was to be used. Dad immediately arranged for a 500 gallon storage tank on to be placed on his farm. This gave us sufficient gasoline to serve dozens of disadvantaged folks.

Regardless of his reason, Dad had a special concern for disadvantaged people. This lesson has been with me all my life. Another lesson, one has responsibilities for one's parents. It is not OK to turn your back on them, or any older person, because it is more convenient to do so. My Dad always held this attitude towards his Mother in spite of the fact she was NOT a pleasant person to be around. He overlooked her disposition and always took care of her physical needs as best he could. This was a most salient life lesson my father taught me.

Fresh meat was also in extremely short supply, if it were purchased through one of the national meat distributors, such as Swift, Armour Cudahy or Oscar Meyer. It was required of all retailers to collect ration stamps for meat, fresh, frozen or otherwise processed sold to the public.

Since there were numerous farms around Odessa producing pork and beef it was possible for Dad to buy livestock directly from a farmer. The government had issued strict price controls, too low for farmers to accept. Dad only needed to pay above the government ceiling to buy all the beef and pork he could sell through the store.

Meat portions were delivered to my Dad's store where they were immediately placed in a refrigerator. This was meat that could be sold without ration stamps, but of necessity it had to be sold at higher than legal prices. This is called commonly referred to as the 'black market'.

As the reader can see here a 'black market' is not always formed for some one to make a huge profit on a war but because a way must be found to provide the essentials to the human population. During times of stress and urgency, government leaders neglect large portions of a population, favoring those who can immediately produce materials of war or have political influence. For many of the rural folk around Odessa, the obvious movement of food produced locally to the people who desperately needed it was essential to the lives of some people.

The need was acute with some families. While we were of necessity charging above legal prices for meat, most folks were very appreciative to be able to purchase meat at any cost.

Something I would like to add here. While it would have been very possible for my father to buy these farm animals, slaughter them and sell the meat at inflated prices to retail groceries in Elmira, Ithaca or Corning, he refused to do that. Dad was concerned only with helping his customers, the people in HIS community. These people, he regarded as friends so he took responsibility for them through difficult times.

With the coming of the war and food scarcity the white tailed deer herd became an attractive source of red meat. As the white tailed deer increased and red meat (beef) became strictly rationed as well as expensive, a growing feeling among land owners developed. Rural folk, had fed and raised deer, hence felt an ownership in them believing they should have priority during harvest time.

As an adolescent I had strong interest in guns and horses. I was constantly hunting small game such as rabbits and squirrels which were always put on the dinner table. I soon came to know several men who were habitually shooting (poaching) deer.

I was soon included in poaching activities. One of the men I worked with owned a pick up truck. When hunting, I stood in the bed behind the cab, with a rifle, another man operated a searchlight, as the truck was driven along back country roads.

The searchlight was an automobile headlight, hand held and wired to the truck battery. A beam of light was swept across the countryside as we drove slowly by.

When the light picked up a deer, it would generally freeze, its eyes shining in the direct beam of light. Deer will typically remain motionless for a few seconds, with its size and sex clearly visible.

Now it was my turn. I had a 30-30 rifle with a scope on it. I could easily hit the deer by using a snap shot even before the truck came to a complete stop. Very few people can master that degree of skill using a rifle or pistol. I prided myself on only using one shot to bring an animal down. Almost always it was a shot centered on the heart.

The deer dropped, dead when it hit the ground. Once the animal was down, men rushed from the truck to the deer and immediately started to dress (clean) it. The truck continued cruising the road watching for a hidden Game Warden. On a quiet, clear night a shot could be heard for several surrounding miles. If all were clear, the truck returned in about 15 minutes to load the dressed carcass that had been dragged to the edge of the road.

The carcass was then loaded onto the truck and driven into a barn. All participants joined in cutting and packaging the meat. With a group of experienced, capable men present, we might take three or four deer in a single night, venison was evenly distributed to each participant.

Before the reader hastens to pass judgment on this activity, remember this was food intended to feed families.

Health and adequate nutrition issues were at stake. Survival in some cases, depended on venison. None was ever offered for sale solely to make money, although at times it seemed an attractive way to gain a few needed dollars. These men were honor bound to use deer only to feed their families.

I became highly skilled with a rifle and developed the skill to "snap shoot" This came about because I developed two personal traits. One, a person needed the obvious familiarity with firearms. Two: also needed, was an attitude of <u>willingness</u> to use the firearm under certain situations.

If I had a fraction of a second to sight a clear target, it was as good as dead, reaction time became extremely rapid, and bullets nearly always hit the point of aim, for a clean kill.

Having these two particular skills were essential later while serving in the military and necessary to save my own life. This was a degree of skill NOT taught by the military at that time. The military barely taught recruits which end of the gun the bullet came out.

While in basic training more time was spent learning to disassemble, clean and reassemble firearms than learning HOW to use them to hit a target. My firearms skills were developed prior to military service.

People tried to find ways legal or not to get food. Almost any method was socially acceptable. Along the shores of Seneca, Cayuga and Keuka Lakes there were several orchards producing Bing and Sour cherries, Alberta peaches and Concord grapes. Usually these crops were sold to canneries where they immediately came under Federal rationing and price control programs. However, sold directly from the farm to the consumer, these regulations could not be policed.

Dad, as well as other grocers, in the area made a practice of going to the orchards with a crew of people to pick the fruit. Dad then took the harvest directly to the store where a load of fresh fruit was sold within a few hours. Peaches were a favorite, as they could be canned easily without sugar for later consumption by a family. Canning was the method generally used in homes to preserve food because freezers were not generally manufactured or available to the home market. Commercial freezer-lockers, available to the public could not be used to preserve illegal venison.

Grapes were included in this practice. Concord grapes were made into wine at any of several wineries in the Finger Lakes area. Families made bulk purchases of available fruit to preserve for home use. Grapes were made into wine or jam, apples into cider or applesauce, peaches were canned.

Military Mobilization

Selective Service (Draft) Boards were organized in 1940, about a year prior to the Pearl Harbor attack. Originally, men selected were to serve one year receiving training only. Since there was no war at the time there were no assignments to combat situations. With the events of Pearl Harbor, Germany's declaration of war on The United States and a declaration of war from President Roosevelt, Congress immediately changed the law, requiring all draftees both those already in military service and those yet to be drafted were to serve for "the duration" in whatever capacity needed. Some early draftees served four to five years of continuous duty, much of it in combat.

The build up of military forces to fight in two theaters of operation, the Pacific and Europe, immediately required large numbers of men to be trained. Every male over the age of 18 was needed in one of these theaters of combat. Almost immediately, skills of all kinds were in demand. The Military had critical use for countless numbers of basic fighting infantry soldiers, mechanics, pilots, truck drivers, cooks, sailors, heavy equipment operators. The list seemed endless. Strong capable leaders, both military and civilian needed to be identified, trained and put into positions of authority.

Young men, still in their teens, were trained to be Officers. Officers Candidate Schools turned teen agers into officers in 90 days. Most draftees were given little choice of assignment they were sent to the branch of the Military where most needed at the time of medical examination and induction.

Service flags began to appear in home windows. A Service Flag was white fabric, about a foot square with a red band about an inch or so wide wrapped completely around four sides, enclosing the white field. Sewn on the white field was a blue star for each member of the family serving in the military. When the blue star was replaced with a gold one it represented a deceased warrior. Service Flags, with blue stars, were soon hanging in the front window of many homes. It was a special day of mourning for the community when a blue star was replaced with gold one.

These mothers came to be known as GOLD STAR MOTHERS and were given a great deal of respect in the community. I remember seeing one Service Flag with five blue stars sewn on the white field. What sacrifices people were making.

Almost every young man taken into the Service left behind loved ones, who might be a parent, sibling, girl friend, wife or children. It made no difference. Although a wife might be pregnant or disabled, needing her husbands attention, never the less, such men were drafted into service as needed.

Elderly parents were left helpless as loving sons who had been making a home for them were brutally torn away. Nearly all families felt intense pain when they were ripped apart. Too many young men, including fathers, never returned home.

Sometimes their remains were never recovered. Sailors killed while on shipboard were buried at sea. Fighting Infantry and Marines were buried near the place they fell. At war's end American military cemeteries existed around the world.

Word of Odessa's first combat casualty came all too soon. Ed Hall, a young man who had been clerking at Washburn's Hardware store was one of the very first to be drafted. He and his girl, Betty the love of his life, were recently married and starting a new life together. She was attending college studying to become a school teacher.

Ed, who clerked at Washburn's Hardware, was well known up and down the streets of Odessa. If one needed any item of hardware, in Odessa, it must be purchased from Washburn's hardware. I became acquainted with Ed while in the store, purchasing .22 caliber ammunition. This bullet was used when hunting small game like woodchucks, rabbits or squirrels. Ed was well versed with guns and ammunition and taught his customers a few of the basics of shooting and safe handling of firearms.

Since Ed and Betty had no children, Ed was among the first to be drafted. It seemed Ed had hardly left home for basic training when his wife, Betty, received a telegram announcing his death.

Given the short time frame, between his departure into military service and his death it was painfully obvious Ed had been sent into combat with little if any training for combat situations.

Very little could be learned about his military service. It is known Ed was sent to Ft. Dix for six weeks of basic training. Six weeks is scarcely the minimum time necessary to train a fighting soldier in the art of war.

New skills and attitudes must be learned. A peaceful civilian must be taught to become a killer. Physical conditioning must also be achieved. Soldiers must become proficient with both his weapon and his job, for military activities are a team effort, not an individual act. With the war just starting, the government was in a hurry to get men into action. Massive numbers of troops were needed in the European and Pacific theaters of operation. It is a fair assumption Ed was sent into battle with practically no training, because there had not been time enough for that. This was a good man, thrown away by our strategists.

While little is known for sure, Ed was part of the invading force landing at Casablanca. He apparently lived through the violence of the invasion. It is thought he was killed later by enemy action either in or near the Kasserine Pass. Since his body was not returned home for burial, Ed must be one of those buried in an unmarked grave in North Africa. Possibly he was blown into pieces by some explosion or Ed might have been taken prisoner and later died while a German POW. Betty never knew.

Merle Joseph graduated Odessa High School June 1942. His father, an elderly widow, operated a small grocery store in Alpine, a village near Odessa. Merle was an average academic student, physically too small for sports like football.

Since he was from outside the village of Odessa, Merle was not known for his student activities. He left school each day riding the school bus, immediately reported to his fathers store in Alpine to do some of the harder manual tasks like stocking shelves and cleaning and improving the appearance of the store.

Upon graduation in June 1942, Merle was immediately taken into the Army Air Force. He was sent to gunner's school and learned to shoot a pair of .50 caliber machine guns from a flying aircraft. Being small in stature, he and the bombardier crawled through the narrow passageway to the stinger (the stinger was the gun position in the tail of a B17), where his assignment was to operate a pair of 50s.

Merle's Squadron, part of the 5th Air Force, was based in North Africa, occupying an airfield captured after defeating the German Africa Corps who retreated from Africa to Italy.

Most missions were carried out were over Italy, not to Germany itself. The Air Force was constantly bombing German positions up and down the Italian peninsula. The scene was very much like the bomb runs B17's were making from England over Germany later in the war.

German fighters were attacking our aircraft during every mission. Merle developed considerable skill with his guns and was credited with knocking down 5 German ME 109's.

Tragically Merle's luck was to run out. On his last mission, Merle's B17 was hit hard by a German fighter. The last radio transmission described the aircraft as going down in flames, no parachutes were observed opening from the falling aircraft. If any remains were found after the crash, the Germans presumably buried them in an unmarked grave somewhere in Italy. The Air Force never released the area where the crash occurred, if it known.

WAR had come to our little community.

Merle's elderly father was crushed with the loss of his only son. His store was located in a dying community, so his business too was failing. He had previously lost his wife through some painful and protracted disease. This was too much for the man to bear. His physical condition failed rapidly. He soon joined his wife and son in death.

My Uncle Fordman Larison (Torp), Dads younger brother was to serve in the combat engineers in the European theater of operations and was badly wounded in the battle for Normandy.

I must digress here for a moment to describe my uncle and his injuries. Torp played the coronet and was apparently very good at it.

Prior to the war he had served several years with the Army Ceremonial Band in Washington D. C. and was one of the few chosen to play at formal State functions.

He earned the privilege of playing taps for burials at Arlington Cemetery. He did this several years prior to the war. Torp never had thoughts about leaving his position with the Ceremonial Band. He intended to retire from the Army.

With the coming of the war, most members of the Ceremonial Band were reassigned into a combat unit. Torp became a combat engineer, a most dangerous assignment. After several months of training in England on June 6, 1944 he took part in the Normandy invasion. Torp was among the first to wade ashore on Omaha Beach. When the ramp dropped, he left the landing craft under intense fire from both artillery and small arms. A hail of machine gun fire raked the inside of the landing craft killing several soldiers before they could charge out. As he dodged his way across that open beach he witnessed friends falling beside around him.

Torp climbed rope ladders to German gun positions with the intent to destroy them. His assignment at the top of the cliffs, was to fight his way close to a pill box (bunker) and toss a satchel charge inside through a gun port. Bunkers housed cannons or rifles that were shelling our ships off shore.

Torp's satchel charge usually silenced that gun. Other infantry GI s completed the destruction with grenades and rifle fire.

A few days later the action on Omaha Beach was followed by further intense combat with the Germans in Normandy, the area known as the "hedgerow country". In Normandy his assignment was to find and defuse German land mines.

After the war my Uncle related some of his experiences to me. He was describing some of the terrible things he had observed. I did not know whether to completely believe the following story.

My uncle was known to be a story teller and inclined to exaggerate. Naturally I thought maybe he was just spinning a yarn. Subsequent events proved his story was all too true and all too common.

My uncle told me of several events occurring during the drive towards Paris while fighting their way through the Normandy peninsula. Our army was moving up fast. The Wehrmacht (German) soldiers were surrendering in large numbers as we overran their positions. Torp described a situation to me where our army did not have food or a place to put prisoners and there were certainly no personnel available to guard them.

Several times Torp was aware that the Lieutenant would order a squad of our Infantry to take a group of Germans back to 'the prison compound" located a few miles to the rear. Well, there was no prison compound. The Lieutenant also ordered the squad to return in ten minutes without prisoners. The intent was obvious as there was no compound for prisoners. My uncle insisted this was almost a daily occurrence.

Much later in my life I watched a television documentary about that particular action in Normandy. The program described the scene I just told you about. The documentary also revealed General Bradley, our Commander of operations in Normandy, in a communication to battle commanders in the front line sent a message to front line commanders, "We don't want to see any of those guys back here", when speaking of German prisoners. Similar events occurred on most of the Pacific Islands, including The Philippines and Okinawa. The Wehrmacht has been documented as having done this at Malmody while defending Germany from the Allied advance at the time if "the Bulge".

Torp also described how German prisoners, even members of the SS, were taken prisoner. When our troops needed to know the disposition of strong points in German defenses during a firefight, prisoners who might have this information were selected and questioned. SS and Officers of course resisted interrogation. However, our soldiers 'squeezed' the prisoner to obtain needed information, using whatever threat or painful method necessary.

If the German continued to resist and subsequently lost his life during the process it was "just tough". Another prisoner who had witnessed the previous scene would be selected for further interrogation.

Usually within a few minutes needed information about the exact position of Wehrmacht troops was forthcoming. Appropriate action could be taken immediately, before the Germans could make troop adjustments. Brutal, yes but these methods saved countless G I's lives.

Torp was seriously injured while handling a fuse removed from a German mine. As he explained, the Germans were constantly changing the design of ordinance. Sometimes this was done to make the ordinance more efficient and sometimes to confuse American soldiers into making a mistake.

Torp was trying to disarm a fuse that appeared slightly different than those seen before when it exploded in his hand. There was severe damage to the tendons in the palm of his right hand the knife he was holding protected the hand from complete destruction.

It took several surgeries to repair the damage, he never recovered full use of his right hand as the finger tendons never healed sufficiently to give him full use. These tendons are critical and must respond perfectly to play the coronet in a professional manner. This devastated Torp, for his life had been performing before live, appreciative, and emotional audiences.

He ultimately developed emotional problems himself. When he was released from a military hospital and reentered civilian life, he was not able to successfully perform a responsible job. For some time he drove a taxi cab around Havre de Grace, Maryland.

Long term emotional problems resulting from the War were not recognized or adequately treated at the time. Torp died at 82, an alcoholic. Poorly controlled diabetes caused the loss of both legs long before his death.

They had been amputated above the knees, rendering Torp wheel chair bound for several years. His was an ugly finish for one who gave so much to his country.

This is another of the hidden costs of war; a price paid the remainder of the soldier's lifetime. Tragedies like this one continue long after the returning smiles have faded, long after parades have passed and flags whipping in the breeze have been taken down and put away, long after war is forgotten.

Next to my Dads farm 'Stormy' Underdown lived with his family, who were dairy farmers. The Underdown farm joined my fathers along one boundary. Stormy, their youngest child, and I hunted woodchucks on our farms during our high school years using .22 caliber rifles with telescopes. Each year, while hunting alone, the two of us were in friendly competition to see who could shoot the most chucks. Actually we were well matched, averaging about 60- 80 woodchucks for each of us every year.

We were both very proficient with the .22 caliber rifle and practiced constantly. No question, neither of us had difficulty hitting a sitting woodchuck in the head at 100 yards.

A high school teacher, Mr. Charles Miller, took responsibility for a rifle club. He was the one who taught several of us the basics of safely handling a firearm and shooting accurately. Mr. Miller would take several of us to the then existing rifle range outside Odessa for lessons several times through the school year. Under his supervision most of us in the club became excellent, accomplished marksmen. I tell this here because these shooting skills were necessary to save both our lives during military service.

Stormy had been raised a Christian in a Methodist family with parents that were very active in church affairs. His father was continuously returned to a Trustees position by voting members of the church.

Like my uncle, Stormy took part in the invasion of Omaha Beach. He was one of the soldiers who swarmed ashore when the ramp on his landing craft dropped. He too ran into a hail of small arms and machine gun fire as his feet touched the beach. Stormy weaved and dodged his way to the cliffs overlooking the landing beach. Since he was such an excellent shot he made his bullets count by aiming directly into the small slit in pillboxes where machine guns were firing from. He admitted to silencing a couple of them. Once Stormy picked a target, it was doomed for his bullets rarely missed.

Stormy was later involved in the hedgerow fighting in Normandy, assigned to an Infantry division. He was usually designated first scout, first squad, and a deadly post. Not many soldiers survive that assignment.

While moving up the Normandy peninsula, Stormy's squad was ordered to approach small French villages that dotted the area. This part of Normandy is also known as the hedgerow country. The hedgerows might be described as piles of rock and dirt dividing cultivated fields reaching around 8-12 feet high, sometimes more and were covered with a forest of brambles, briar's and underbrush. These were ideal places for a Wehrmacht soldier to hide, set up machine gun or mortar positions to fire on approaching Americans.

When the Americans entered a village one could be sure there would be German snipers in the bell towers or other high places. These snipers could shoot down on the Americans, picking off key members of a squad, like the officers, non-com's or those carrying particular arms like flame throwers or BARs (Browning Automatic Rifle) or mortars.

After several close encounters with hidden snipers, Stormy decided two could play that game. He stayed out of town, hiding in a nearby hedgerow keeping well hidden, yet in a position where he could shoot into the town.

He was looking for German snipers hiding in church bell towers or other high places. Earlier I mentioned, Stormy had no trouble hitting a woodchuck in the head at 100 yards with his .22 cal rifle. Likewise, it was no problem for him to hit a German in the head at 500 yards with his .30 caliber Garand rifle.

When telling me this story Stormy mentioned that his Garand would only fire single a shot. That gun was made so it can be switched to fire semi automatic. I asked him why he didn't drop the malfunctioning rifle and pick up another that was working properly. Stormy replied.

My rifle had never been used rapid fire (semi-automatic); the rifling inside barrel had never been burned, retaining its original accuracy.

He explained, barrels on most Garands had been burned out as a result of rapid fire. Continuous rapid fire heats a barrel to the melting point. After this kind of use the Garand will not shoot accurately at long ranges. Stormy said my gun would put a bullet right where it was wanted. One bullet, one enemy sniper eliminated.

To effectively do his job as a sniper, it was necessary to find a target, watch him, sometimes for hours, until the proper opportunity came to make the shot. In this way he developed personal contact with his victim. Killing became very personal. In effect, Stormy knew he had killed a real person and this haunted him.

While Stormy never received a bleeding wound, so could not receive a Purple Heart, he carried deep emotional wounds the remainder of his life. He could never forgive himself for 'murdering' other men. I really don't know how many Germans he shot from this position, but I suspect there were several Germans killed during his time of combat. Killing was certainly not a single event. Stormy was never offered, sought or was given any psychological help with the guilty feelings plaguing him. As he aged, he became increasingly unbalanced, eventfully becoming a recluse.

Jim Cunningham, another Odessa man, had been an Officer with the Border Patrol stationed at Niagara Falls, New York prior to the war. With the bombing at Pearl Harbor, Jim was among the first to volunteer for active military duty. He soon earned a commission, becoming an Officer in the Marine Corps.

Jim's first action was on the island of Tarawa. While we conquered the Island, defeating the Japs, the invasion of this island was nearly a disaster for the American Army as a result of poor planning. It seemed everything went wrong. The tide was too low. Jap fire more intense than was predicted. Marines were slaughtered while pinned behind a sea wall, resulting casualties were much higher than expected.

Jim was one of those Marines pinned down behind the sea wall where so many of our men died. Since he was an Officer it was up to him to almost force his men over the top of the sea wall to advance further into the island. The Marines took the small island but at a heavy price in causalities.

After the battle at Tarawa he was made a Captain, responsible for hundreds of combat Marines. He was to be part of the invasion force into many of the Pacific Islands. It became his unpleasant duty to order Marines into harms way where many of them were killed by enemy action. He served four years in this capacity.

The last action Jim was part of was the invasion of Iwo Jima. He ordered platoon after platoon into what amounted to a meat grinder, Jim, ordered hundreds of men into almost certain death.

I understood unofficially 75% of the men he ordered into battle never returned. They were casualties, either killed, wounded or missing. That is a terrible burden to ask one man to endure.

When Jim returned home after the war ended he too was a changed person. No longer was he the decisive leader of men, no longer could he make simple decisions. Jim had become defensive and withdrawn, socializing with others grew increasingly difficult making it nearly impossible.

Jim experienced almost weekly flash backs. The hallucination was always the same. He could see men passing by his viewpoint. In his hallucinations it was possible for Jim to see the faces of men he recognized staring at him as they floated by. These images plagued Jim for the remainder of his life.

He was never able to return to the Border Patrol, as any position giving orders to others was out of the question for Jim Cunningham. He hopelessly sat out the remainder of his life in his home with wife Janet, subsisting on a Veterans Administration disability pension.

This strong man, who had accomplished so much, became an alcoholic eventually taking his own life, unable to overcome the guilt war had produced in him. His wife, Janet, who at the beginning of the war a healthy vibrant woman, very much in love with her husband nursed Jim through his agony while it was taking a toll on her. She too died at an early age, completely spent both physically and mentally.

George Darfler graduated The College of Agriculture at Cornell University, June 1941. George had participated in the Reserve Officers Training Corps (ROTC) while a student at Cornell University. Immediately after graduation, June 1941, George entered the Army. Since he had taken several engineering courses at Cornell, he was assigned to the combat engineers, which was only a skeleton branch of the Military at the time. Our leaders, sensing a major war was coming started preparing and enlarging military units at least a year before hostilities began.

George was among those landing on North Africa at Casablanca. Now a Captain, George headed a unit of combat engineers. During one engagement with the Wehrmacht, George was assigned to knock out German Panzers (tanks) any way possible. One of the techniques used was to send out a patrol of about 8 or 10 soldiers carrying several 5 gallon cans filled with gasoline. They entered German lines, found a panzer that had been dug in, half buried in sand. The retreating Wehrmacht was out of gasoline, so dug in their Panzers and began using them as artillery. Georges' men carefully sat a Jerry can of gasoline on the panzer, stepped back a few feet and shot the gasoline can with incendiary bullets. Burning gasoline instantly set the panzer on fire.

After several successful missions Germans learned how to cope. When metal was heard to strike metal, the sound immediately alerted the Germans inside, knowing what was to follow they came rushing out of the panzer, instantly becoming easy targets for George's men surrounding the panzer. When shooting stopped and the German's killed, explosive charges were placed inside the panzer.

North Africa was followed by Sicily, then the peninsula of Italy. George trained his men to build bridges. After the bombardment accompanying invasions of Italy at Anzio and Salerno most bridges had been knocked down. The Wehrmacht made a practice of destroying bridges when retreating in an attempt to slow the advancement of U.S. troops.

Captain Darfler used a series of pontoons to support bridges that could hastily be thrown across a river so the U.S. Army could continue its momentum, plunging deep into enemy held territory. It was imperative, the movement of heavy equipment, artillery, tanks, trucks and hundreds of soldiers cross rivers or wetlands rapidly.

Temporary bridges were made of a series of inflatable rafts, cabled together side by side across a body of water. These rafts or pontoons looked very much like Zodiac boats used today. Across the tops of the pontoons steel strips, wide enough for truck tires or tank treads, were hooked together forming twin tracks. Once constructed, infantry soldiers and heavy equipment including tanks could rapidly move across. If the enemy destroyed one of these bridges it was quickly rebuilt.

George had mastered details of exactly how to "throw" a bridge within a few hours. His special knowledge was necessary and a team of trained soldier engineers were essential to achieve desired speed in the placement of a bridge. Much of this work was accomplished while troops were under enemy fire, causalities were always high. Behind the battle line, continuous training of replacements engineers was of necessity a continuous activity.

Each engineer assigned to bridge building must know exactly what his job was and how to perform it quickly and efficiently. Speed needed to get the bridges operational was essential. One technique used was for the U.S. artillery and or Air force hit the Wehrmacht simultaneously with everything available.

During this heavy bombardment, German soldiers were forced to hide in their foxholes reducing their fire. During this lull in return fire, George and his engineers 'threw' the bridge across the water. When the bombardment ended U.S. troops and equipment were already across the water.

Somehow George missed the invasion of Omaha or Utah beaches. He continued his specialty, building bridges. A year later when the fierce action at the Remagan bridge subsided it was apparent the original bridge across the Rhine would soon collapse as it had been so severely damaged while beating the Wehrmacht back. Captain Darfler, using his special skills put a pontoon bridge across the Rhine alongside the original structure days before it too collapsed, the result of artillery shells.

At wars end, George returned home having completed five years of almost continuous combat. He was never wounded or injured in any way, returning home in healthy condition although underweight and in need of a considerable amount of sleep. As he described the experience, "it seemed my own army was trying to get me killed, for every difficult and dangerous job was given me time after time.

Not once or twice did he have very close escapes from death or injury, it was almost a daily occurrence. Possessing special skills, George was in constant demand to build bridges while our troops were pursuing the Wehrmacht.

When George returned home he signed up to continue in the Reserves. Since he held the permanent rank of Captain he felt this could be a nice income with possible retirement benefits and very little continuous service obligations.

Then, another war, Korea. Again the call went out for Captain Darfler. By now George had other interests, including a lovely wife and children. Further active service did not appeal, but it was no use. He had signed an enlistment in the Army Reserves so was required to serve another two years, as a combat engineer. Again, he was placed in combat situations demanding much from this man.

Unfortunately George died soon after returning from Korea. It was such a tragedy, this man who had survived two wars without so much as a scratch on his body died from a heart attack.

After a careful autopsy, the doctor who had been caring for George declared his heart failed him possibly because of the continuous stress he had sustained for so long a period of time. Korea was simply too much for him to survive, his heart was worn out.

He was barely into his 40's when he expired, leaving his wife and three children. Since George had never applied to The Veterans Administration for medical care, the government refused to accept his death was directly caused by his military service. His widow could never collect a Veterans Administration survivor's pension.

Allen Couch graduated Odessa High School June 1942. He could be described as an outgoing personality, full of life, very social, from a family of well respected people who were leaders in the community of Odessa.

Allen also played six man football and was along distance runner in track. A handsome boy, it seemed he always had a good looking girl hanging on his arm.

After High School graduation, he was immediately inducted into the Army. Allen did spend some time State side receiving considerable training in Infantry tactics. He volunteered for The Airborne, receiving more specialized training. For a time, he served as an instructor in the 101st Airborne before being deployed to the European theater of Operations.

Allen was one of the parachutists dropped in France at the time of the Normandy invasion. Many of those soldiers missed their assigned drop zones by several miles.

He, with a few comrades, fought small skirmishes for some time before they were reunited with their unit. While Allen was not injured physically, he felt he had failed his unit for he spent much of the time hiding from, not engaging the enemy. He experienced guilt as his buddies died during the ordeal and his unit struggled against the Wehrmacht.

The next jump he was willing to talk about was the operation into Belgium code named "Market Garden". Again he missed his assigned drop zone. This time he was unsuccessful in avoiding German soldiers. He was taken prisoner. The Germans thought Allen could tell them something about the "Market Garden" operation. They started to question, "squeeze" him about objectives, strength and the position of U.S. and British troops. As Allen remarked to me. Hell, I didn't have answers to their questions, but couldn't convince them I was really unable to answer.

To squeeze Allen for information, the Germans tied him in a painful position, placed him on a table, followed by placing lighted cigarettes to his bare feet and legs. There were additional beatings. It didn't last long as the battle situation was so fluid whatever Allen might have known was irrelevant within a few hours.

When Allen returned home at war's end, his social skills were noticeably missing. He refused to talk about his experiences in Europe. I knew him some time before he was willing to reveal anything about what happened to him while a POW. I suspect he seldom shared those experiences with another person.

Now introverted, underweight and bent in stature, he looked much shorter than his previous posture. His legs, including his feet and legs were horribly scarred. It was hard to understand how a lighted cigarette could do so much damage. To my knowledge, Allen's family apparently did not understand what had happened to their son so did not press the Veterans Administration for medical assistance. He never turned to alcohol or drugs, he simply faded away, it is doubtful he ever held a job. A bright life, lost.

Soldiers, disabled during WWII were soon viewed as burdens to society and found themselves in constant battles with the government that caused the disabilities. To simply receive adequate living expenses and/or medical care became an ongoing battle with government bureaucracy.

The government refused to recognize developing emotional problems as a result of WWII and make it possible for The Veterans Administration to care for people these situations.

My Service Begins

Early summer, 1945, it was evident the War with Japan was coming to a violent climax. Germany had surrendered, troops who had been fighting in Europe were being transported to the Pacific Theater of operations. An invasion of main land Japan was carefully considered by President Truman and our military leaders.

With the recent experience of Iwo Jima, Okinawa and the Kamikazes we could expect that an invasion of Japan would require out troops to fight not only every soldier, but civilians. We would be fighting and killing men, women and children, the Japanese were so determined to keep us out of Japan. Losses on both sides were estimated to be in the multiple millions of people. Japan was now fighting for their homeland.

The U.S. Military had drawn plans (Called Operation Downfall) that would put every available serviceman in a mass invasion of the southernmost Japanese Island, Kyushu. Later invasions were planned on northern island of Honshu, near Tokyo. Tensions in Americans raged high as any serviceman taking part in an invasion could expect the worst possible experience, with death a strong possibility.

With this scenario under consideration, is it any wonder the American population cheered when the atom bomb was dropped bringing about an immediate surrender of Japan. It was over, enough, finally an end to the slaughter.

Since I was very young and not world wise, I was eager to enter the Service and fight those who would harm us Americans. Dad, being wiser than I, had lived through the experience of WWI as a teen ager. He knew of the horrors of war. He understood men died and or returned home disabled, receiving little or no consideration from the government, and uninvolved people soon forget... Dad described it at "used and then thrown away, like old shoes".

Veterans of WWI were given very little recognition of their service. The Government did little to care for the wounded. NO Veterans Administration to help veterans existed. Disability pensions were meager. Medical care was less than adequate or non existent. Disabled soldiers were left with few resources.

Dad also spoke of the poor treatment returning soldiers from that war were subjected to by the society at that time. General MacArthur, who had been one of their own during WWI, led the Military in the brutal quelling of ex soldiers when they marched on Washington D.C. in 1933 demonstrating and asking for a soldier's bonus.

This event took place during the Depression when many, including discharged soldiers, were out of work, out of money, with no way to feed themselves or family.

It was common for unemployed to seek a meal from food kitchens operated by The Salvation Army. Yes, Dad knew of these things and did not encourage his son to get involved if an alternative could be found.

In early 1945, Selective Service (the draft board) ordered me to Syracuse, be given a physical exam, if I passed, was this to be my time in hell or a lark? I vowed to endure this with little pain or strain and seek positive experiences whenever possible.

The physical given at the Armory in Syracuse was quite thorough but the outcome predictable. Immediately after the physical exam all draftees were loaded on waiting trains to be delivered immediately via non stop Pullman to the reception center at Fort Dix, New Jersey. Dix, a long established facility, was laid out beautifully with nice barracks.

At Fort Dix, draftees were introduced to The Military. The required shots were first on the agenda. They must have thought the American lad was a pincushion. If memory serves me correctly, there were seven shots given within a few minutes?

One was hit on both arms at once, not by a nurse or corpsman but a grunt soldier like myself who obviously had no idea what he was doing or why. No wonder some recruits became sick.

Next it was a uniform. Uniforms never fit, with little effort given to fit clothing to the recruit. If too large, you were told, "you will put on weight". If the uniform handed you were too small the comment was "you will loose weight". None of us felt uniforms as issued were something to be worn with pride. They were nothing more than shapeless rags.

The only item issued that received any care was the shoes. Our feet were measured carefully with and without extra weight. An army marches on its feet remember? Boots must fit or the soldier will fall out. There were many miles to be hiked and many packs to be carried. We were issued Combat Boots and a pair of what one might call hiking boots. Both were a joy to wear.

We were issued a combat knife. What a weapon, the blade was about 8 inches long with serrated back edge. Probably it was an effective weapon for hand to hand combat should a soldier have the need, but why were we issued that thing at this time when the war was history, actual fighting was supposed to be past?

Let' don't forget the helmet liner. This was the shell fitted on the head, under the helmet. When going into action the steel helmet, commonly called a "piss pot" would fit down over the liner. Those things weighed about 8 pounds. Wearing them a few hours almost guaranteed a headache. Fortunately for us, only liners, not the steel helmets, were issued.

Finally, "duffel" or a "barracks bag" was issued. An empty barracks bag was circular like an empty pipe, about three feet six inches long with a width of about two foot six inches.

We were to put all government issued clothing and equipment inside them. Believe me, issued equipment packed those barracks bags full, loaded bags were heavy with little space remaining to squeeze in personal belongings. Stenciled on the outside of the bag were the soldier's name, service number and branch of service.

The barracks bag was to be my constant companion from that day until the date of discharge.

That fully packed bag, was loaded on trucks many times during my time in the Military, carried to one location or another. When unloaded, several trucks might arrive at a destination simultaneously, the bags thrown from the trucks into one great big pile. It was each soldier's responsibility to find his own. One could spend hours looking for your own barracks bag.

There was one final item to be attended to, that is the dog tags. These were made of aluminum and had stamped the soldiers name, serial number, religious preference and blood type. There were two of them on a chain to be worn around the neck.

Training for Service

It took about a week for the incoming processing to be completed. We then boarded a railroad train waiting on the siding for us. We were to travel first class in a Pullman car destined for Kessler Field, (Near Biloxi) Mississippi. After boarding the train we were told we had been assigned to the Army Air Force and were going to Kessler Field for basic training. We were riding in Pullman cars complete with a porter who took down the bunks each evening, made them up, preparing them as if we soldiers were first class passengers. In the morning he took away all bedding and returned the bunk to its position along the ceiling. Was this to be the Army Air Force? Well not quite.

Valuable lessons about living with men were soon learned. Take care of yourself, watch what you say, leave few belongings unattended and never gamble especially with cards or dice. Hustler's, con artists and cheats were everywhere. It was impossible for me to understand the methods used but the results of cheating by deception were evident, even to one not versed with gambling. One innocent looking kid would get exactly what he wanted with every throw of the dice or every turn of a card. To me this seemed like too much for coincidence. Hundreds of dollars changed hands every day.

Apparently our arrival at Keesler Field was not expected. It was around midnight when the Pullmans were parked on a siding inside the gates of Kessler Field. We recruits were unloaded, the train immediately pulled out.

We were left standing in a field with no one present to take charge and direct us to a reception area. I MEAN THERE WERE NO DRILL INSTRUCTORS OR OFFICERS PRESENT. It was hours after daylight before the Command at Kessler took any action to move us to an available barracks.

The shooting war was over, apparently the military command had little interest in giving us real basic training. Several days passed before training schedules were developed. We were to attend classes about six hours a day, just like High School.

The schedule was set up similar to those used in High School where students move from one classroom to another for instruction. Scheduled were fifty minutes of instruction, 10 minutes to march from place to place. Classrooms were under a shade tree or on a drill field, rarely inside a building where we might be protected from the hot sun or rain.

Classes consisted of close order drill, physical training (calisthenics), rifle range, explosives, chemical warfare, obstacle course military law, military protocol (saluting an officer, a salute must be executed properly, and always say "Sir" when addressing Officers). Let us don't forget, much detail regarding the wearing of the uniform. I seem to remember considerable time was spent instructing us how and when to wear the Class A uniform.

Actually the only time I wore my Class A uniform was when I went home for a ten day leave. In those days the uniform must be worn at all times, even if the soldier were on a pass or on official leave. NO CIVILIAN ATTIRE COULD BE WORN AT ANY TIME while on active duty.

Our barracks was a two storied building positioned very close to the end of the runway. Each barracks was supplied with cots for 30 men on each floor. This made a total of about 60 men per barracks. Since Kessler Field was an Air Force base, aircraft were constantly practicing touch and go landings passing directly overhead while approaching to the runway. When a "heavy" passed overhead the entire building rattled. Want some undisturbed sleep? Ha Ha.

There were no trained or experienced DI's (Drill Instructors). We were told to elect one of our own. His only responsibility was to get the class of recruits (now called a flight) to the proper class on time. He was also a kid off the streets of New York, so obviously knew nothing about training men to be soldiers. The poor kid we "elected" to be our D.I. was given the "honor" because no one else would take the responsibility.

The poor boy knew no more close order drill that we did, but was required to march us in formation and in step whenever going to classes or the mess hall. Obviously he needed the cooperation and good will of all in the flight to accomplish anything. We soon learned marching songs, none fit to repeat, the raunchier, the better.

I buddied up with a boy, also 18 years old, Richard Mercer from Columbus, Ohio. He was of slight build, extremely intelligent and willing, but he had a small light physical stature that made it very difficult for him to carry a pack or march.

Dick also had varicose veins in his legs. Walking and marching any distance was very painful for him. I liked the boy a lot but he sure needed someone to look out for him.

There were a few days of learning to march, how to make our beds to pass inspection, military customs, saluting all officers even if he is out of uniform. Be sure to include a very loud 'Sir' with each sentence. Military law seemed to be a stream of threats. Soldiers could be Court Marshaled or given company punishment for any offense dreamed up by some Officer and would be represented in court only by another Officer. Were we to become soldiers as the result of constant threats?

It was soon perceived by recruits, Court Marshals were simply Kangaroo Courts, as soldiers were given little, if any opportunity to defend themselves. We were never given a copy or allowed to see any printed form of 'regulations'. They could be whatever any Officer wanted them to be. Streams of constant threats do not develop trust between Officers and enlisted men.

Let me tell you about the day I learned about the "short arm" inspection. It was a hot day in Louisiana. The nice Lieutenant came into the barracks and gave us orders for the "uniform of the day". It was to be boots, helmet liner and raincoat only, no under shorts and no jock straps. It was a hot sunny sun day in Louisiana with a sun blazing overhead. Is the Military brass nuts?

We were marched down the street in formation hup two three four, did we ever feel silly, that is wearing only a poncho in the hot sun. Inside a supply warehouse were a cluster of officers, all wearing Medical insignia that decorated the uniforms of both male and female. Let's don't forget the non-com's, also present, constantly serving Commissioned Officers. We were to step up in front of a table, open up the raincoat and expose our private parts to some female. This was called a "short arm inspection".

It seemed the military could have at least assigned a man to do the visual inspection. Perhaps since this was our first "short arm" it was intended to be embarrassing.

The nice nurse (female) said and I quote said "Skin it back, squeeze it down and turn it over soldier". All WWII soldiers, airmen, sailors and marines endured this indignity at least once a month as long as they were in the service.

What is a military Mickey Mouse film one might ask? A Mickey Mouse film is about sex education, especially the venereal disease (VD) issues. These films, and there were several of them, showed the terrible consequences of contracting any of these diseases. Beautiful young women using all their female allure were portrayed to show exactly how insidious VD can be. These beautiful, virgin like women, it was claimed could infect a soldier with horrible sexual conditions. The first time one saw one of these presentations, a soldier might walk out of the theater vowing never to touch any woman again.

However that attitude soon passed as shock value became an 'old story'.

All soldiers were to see one of these films at least once a month as long as they were in the service. They became a standard joke, referred to as "Mickey Mouse films." After the initial shock they were good for a laugh if nothing else. Besides if the ordeal was ordered, a soldier might be excused from some other distasteful job.

We were told about condoms, and pro stations. Condoms were always available from any of several sources. At the gate the guard assigned to the post was authorized to distribute them to anyone who asked. We were certainly encouraged to keep them in our possession. In addition there was a 'pro kit.'

A 'pro kit' consisted of two condoms, a piece of coarsely woven cloth, impregnated with yellow soap, a tube of antiseptic also containing antibiotic and sulfa tablets.

After visiting a female, soldiers were to remove the condom without touching the outside of it, scrub himself from knees to navel with the yellow soap and squeeze the contents of the tube into the urethra and work the antibiotic upwards towards the bladder. A pro station was located just inside the gates on all military installations. If a soldier came into a base inebriated, pro treatment was mandatory. Military Police enforced the regulation, making sure of compliance. These were, of course, measures to control venereal disease that, if not controlled, can disable an army. In a pro-station" the soldier was to use a pro kit while under the watchful eye of a Corpsman.

If a soldier signed in at a pro station upon return from off base, there would be no charges should he later develop a social disease. VD was treated without questions being asked. While overseas, we were advised if a soldier had a female in mind, he could bring her to the pro station and she would be checked by Medics. If she were infected, she was offered a course of antibiotics. The soldier was then advised of the situation. If a soldier did not report to a pro station and later developed a venereal disease, he could be give a less than Honorable Discharge.

Do not forget, close order drill, lots of that. The only physical activity that bothered me was double timing. I could run OK but running to the cadence required in double time threw my timing off so much I just could not stay with it. We ran the obstacle course but there were no time limits. This was Boy Scout Camp all over again, for me, I was having a great time.

Classes for some subjects were in the field where we were to have some exposure in learning a few combat skills. At one training location, we were given instructions regarding the technique for throwing hand grenades.

We watched but did not actually throw exploding shrapnel or phosphorous grenades. We received a whiff of tear gas and were introduced to the use of 1\4 lb blocks of TNT with detonators and some instructions in setting up booby traps.

The purpose I suppose was to develop killer soldiers who could handle firearms. However the actual time spent on the rifle range was far too short for anyone to actually learn to use these weapons.

Very few recruits actually knew how to shoot a firearm and our basic training did nothing to remedy that condition. We were instructed and actually shot a .30 caliber carbine. There were no Garand rifles present in fact I never saw one while in service.

The rifle range, ah yes, that place was another adventure. At the end of this training joke which actually consisted of a couple of days on the rifle range. Most time being spent with instructions regarding how to hold the firearm, where the trigger was and how to squeeze. It seemed most time was spent learning how to disassemble and clean the hardware, not actually shooting it. Little ammunition was actually fired, certainly not enough to develop any degree of proficiency with any of the firearms we handled.

After a minimal amount of instruction and shooting less than a dozen shots at that bed sheet, we were to shoot for qualification. This was to be timed rapid fire at the same bed sheet.

The carbine is a nice little gas operated, .30-caliber semi automatic rifle operating both single and semi automatic fire at the flick of a switch. There was a relatively small powder charge behind the bullet so it did not pack a heavy punch. However, as I was to learn later, power enough to kill a man at close range and deadly accurate at ranges of less than a hundred yards.

It was difficult to judge just how accurate the carbine was. At 200 yards we had been given a bed sheet to serve as a target. It had a two-foot black bull in the center for aiming. An experienced woodchuck hunter like me could make a two inch group in the center out of a target at 100 yards.

With the carbine, I could not shoot as tight a group of hits on a one hundred yard target as I could with my .22 caliber rifle at home.

The groupings on the target at 200 yards seemed to be quite large, not tight enough to make a sure kill at that range. In my judgment, the carbine was a rapid fire short range firearm, not suitable for long range precise shooting. Possibly at less than 50 yards my groupings could be about the size of a half dollar if fired with the selector, set on single shot.

The gun I was given would not securely hold the clip when fired. When I touched off a shot the relatively light recoil caused the clip to drop out of the weapon. I would then pick it up, replace it in the weapon, work the action and squeeze off another shot. Believe it or not, the Army Air Force said I qualified for rapid fire. It even says so on my official discharge.

There was one important thing learned while on the range. One could put a condom over the end of the barrel of a firearm to keep dirt and water from entering.

The obstacle course was another training joke. It seemed like Boy Scout camp. The course was nothing, a few jumps, and swing across a mud hole on a rope, crawl under a few logs and vault over a four foot wall. Simple as it was, the city boys could not master the course. It was pitiful to watch them. Most could simply not traverse the course successfully. Probably it was more a lack of self discipline rather than physical ability that prevented them from being successful at most any physical requirement. The city boys did not qualify for this simple course, they quit almost before starting. Am I destined to serve with these guys??.

With the coming of December, it was our turn to go into the southern pine forest. Field maneuvers it was called. We were marched several miles in a warm December sun in woolen over coats from Kessler Field to a long leaf pine grove. Here we erected our shelter halves and made camp.

The temperature pattern in Mississippi in December is mild by Yankee standards. During the day the air is pleasant with temperatures in the 50s and 60s. The temperatures were much too warm for winter wear overcoats, nights could be much cooler, with frost possible. That coat, which seemed like overkill during the day, was most welcome during those very, uncomfortable cold, damp nights. With the continuous high humidity, the cold can be most penetrating and uncomfortable although air temperatures might remain well above freezing.

The military issued each soldier a shelter half. (half a pup tent) The idea, a soldier was to buddy up with another, snap two shelter halves together assembling a pup tent. I had buddied up with my friend Richard Mercer.

Two other airmen joined, we four benefited when we snapped our four shelter halves together. The four shelter halves when snapped together facing each other formed a solid wall protecting those inside from outside temperatures. After entry the all openings could be closed from the inside.

In southern pine forests there are plenty of fallen pine needles available to make a soft bed. For all night comfort the reclining body needs to be insulated from the damp ground. I told my tent mates to gather an abundance of pine needles, making a deep bed of them inside the tent.

The military issued two wool blankets, no nice warm sleeping bags. First spread out the ponchos, placing them on top of the pine needles (fabric on one side, rubber on the reverse side) to keep dampness from coming up from the ground. The best use for two blankets is to roll up in them so your body is surrounded then cover yourself with your overcoat. Dry socks carried in an outside pocket during the day are essential to keep feet warm and dry at night.

Mercer rolled in his two blankets with his overcoat wrapped around the outside. He told me sleep was warm and dry every night. The tent, being draft free with a deep bed of pine needles, was cozy and comfortable inside.

Another day brought another exercise. This was to be with tear gas. We were taken into an open field, where there was very little breeze blowing. Several canisters of tear gas were popped at one time. A gentle breeze swept the gas towards us. There was not concentration enough to distress a person, but it could sure make a person uncomfortable. The eyes teared and swelled up, the stomach rebelled and lungs burned. This was to give us some idea of what tear gas is and how it affects people. Since we were in the open with a gentle breeze, the lesson ended when the gas soon blew away.

My first lesson in 'do not trust what an Officer verbalizes'

Several recruits had been selected to serve our first duty as interior guard or sentries. We were taken to a Provost Marshall's office where an MP Captain began to lecture us on how to be an effective sentry or interior guard.

He was very emphatic when he said "as MP's you have the right to shoot to kill to protect the property you were sent to guard or to protect yourself". Regardless of the Officers rank I mentally questioned the truth of that statement.

With the instructions "shoot to kill," I was issued a .45 cal 'grease gun' with an empty clip. What did they expect me to accomplish with an empty gun? The grease gun is a cheaply made machine gun, a 'throw away' actually. The grease gun looks threatening but of course is powerless when not loaded. The truck dropped me in front of a PX late at night. Sentries were posted to each PX to stop burglaries. We were instructed to walk a post completely around the PX.

It seemed the Command had deliberately put this group of recruits in deadly dangerous situations where there was no way to either accomplish our assigned mission or defend ourselves. Grey decided to stay out of the way, lurking in dark shadows. It seemed obvious Command had NO interest in either protecting property or in the welfare of soldiers placed in "harms way".

Unfortunately, that same night another recruit was sent to guard an ammunition storage area; again the soldier was given the assignment with an empty weapon. He was sent to a remote post where, during the night, he was assaulted and badly injured, enough to require hospitalization, when someone came to steal ammunition. That boy had no means of communication as no radio or telephone equipment was made available to him. He lay on the ground, beaten and injured until his relief arrived the following morning.

Officers making such ludicrous assignments should be severely disciplined for putting a soldier in harms way, with no way to defend himself or carry out his assignment. Obviously some Officers were incompetent and uncaring for men under their command.

Not all new draftees had repairable physical problems. One Sunday morning while walking past the rear steps of a nearby barracks a soldier came out the door and immediately had a grand mal epileptic seizure.

He was standing at the top of a moderately long stairway when he fell headlong down cement steps hitting face first on the concrete apron below. He lay on the ground twitching and bleeding from his face.

I was momentarily paralyzed with indecision having never seen anything like that. What to do? Who to call for help? What was happening to him? How badly was he injured? MP's soon arrived, gave the patient a cursory examination then put the stricken soldier in a jeep and drove him to the hospital on base. It was only a few hours later when an MP came to the barracks and collected the soldiers clothing and equipment. No further information was made available to us. Almost certainly the boy was given a Medical Discharge and sent home.

Early in the game it became apparent some guys tried ways to beat the "system" There was one lad who had been classified Limited Service (no combat assignment) because he had a ruptured eardrum. Our barracks was situated under the glide path for the runway. Day or night, when a large airplane, like a B17, C47 or PBY (amphibian aircraft with the capability to land on either land or water) passed overhead, this soldier had a routine he would perform. He started screaming "the sound of the heavy motors pained his ears and penetrated deep into his brains".

In all fairness, I must add, it was common practice for pilots to shoot landings. It was annoying when they were practicing touch and go landings. A plane would approach the landing strip, come down the glide path directly over our barracks, touch the runway, gun the engines and take off again. The pilot would circle and make another approach. Some might be B17, B24, PBY or C46 or C47 in addition to numerous fighter aircraft. It could be a noisy place for several aircraft might be practicing simultaneously.

This soldier kept up his act until he was sent to the hospital for evaluation. After several weeks, while waving his Medical Discharge over his head, he returned to the barracks to say goodbye. He had successfully postured and was issued a Medical Discharge.

The military agreed to provide him with transportation home. He had the option of taking a train, bus or going on space available aircraft.

His destination, New York City. He opted to fly from Kessler Field to New York's La Guardia airfield. Several of his friends went with him to Operations, watched him get on a B17 (a B17 is a heavy bomber with four motors) and take off for home, a flight of 1500 miles. Couldn't stand the sound of heavy motors, yes indeed.

After several weeks of the basic training nonsense, it was time to be shipped to some further training facility. I was talking to the nice Lieutenant who certainly had no experience or interest in fitting square pegs in square holes or round pegs into round holes.

He was simply filling a list of training school openings. Why bother with this charade of pretend interviews? He asked me what work experience I had, well, at my tender age, very little. I did mention that I had cut some meat in my father's store. "You will make a fine cook" was his response. He had a long list of schools needing cooks.

My next assignment was, at Scott Field near Belleville, Illinois, for Cook and Bakers School. This was to be another joke. For the following 6 weeks there were only continuous KP assignments in one mess hall or another on Scott Field Air Force base.

At Scott Field there was never any attempt at instruction or the procedures to be followed in the kitchen. There was the clipper (a type of automatic dishwasher) where aluminum serving trays were washed and scalded. All trays needed to be run through the 'clipper' after each meal, steam kettles must be scrubbed by hand, floors to be polished and potatoes to be peeled, turkeys to be disassembled into bite sized pieces. Meals were served by hand to individual soldiers coming down the line.

The only useful memory carried away from that experience, I learned to strip the meat from a freshly roasted turkey. No knives were used, the meat simply pulled from the bone and shredded into bite size bits. OK some acquired skill learned, hopefully it will come in handy some day. When we "graduated" we were given the MOS number on our records that stated we were qualified army cooks.

A Sea Voyage

Immediately after completing "Cook and Bakers School" we were shipped to San Francisco where we were to board a troop ship, bound for some overseas assignment. This included a train ride in a 40 or 8 railroad car from Scott Field to Camp Stoneman, located near Pittsburgh, California, 40 or 8's were very primitive, no kitchen and no latrines. The train ride was followed by a short boat trip down the Sacramento River to the docks lined around San Francisco Bay. The excursion boat headed straight to the troop carrier, "Seacat". We disembarked onto the dock nearest our transportation.

A line of us lucky soldiers were waiting at the foot of the gangplank to board the Seacat, the troop ship that was to take us where we did not know. At Stoneman, the Military had issued us clothing, for both the arctic and the tropics. This equipment filled another barracks bag. We now had two of them to carry around. One could only wonder at the logic of this. There was no need for secrecy or deception since the war was over. Why require us to carry clothing we were not to use. As we boarded the Seacat none of us had any idea where we were bound or for how long.

Again a nice Lieutenant addressed us from the gang plank. We did not have to get on this ship, he claimed. If anyone would enlist, he could immediately go home for thirty days for each year of a signed enlistment. Sign up for three years and it was 90 days home, immediately. A <u>bonus</u> furlough, guaranteed.

This was promised to be a bonus furlough and not count against the annual thirty day furlough time granted to all military personnel. This sounded good, but even as a young lad I was aware the politicians were posturing about "getting the boys home". Draftees had this advantage, however slight it might be.

Once a man enlisted, that degree of protection was immediately lost because he became a volunteer, outside a politician's interest.

The actual situation: By voluntarily enlisting, a soldier became a "volunteer" in the regular Army, thus loosing the voice of anyone trying to get him home. Although the destination and the time to be served was still unknown to me the proposition did not ring as a good deal for this soldier who just really wanted to return home as soon as possible. With little consideration for the offer, I lifted both my barracks bags and walked up the gang plank.

That offer proved to be a great big lie. The boys who took the offer soon found themselves at an overseas post for the entire length of their enlistment, without any furlough time. "You had your furlough", was the official response to a request. There were guys in my Squadron, required to spend 33 continuous months in the Philippines. An important lesson to be remembered, the government, especially the military, is not to be trusted, under any circumstances their word is, more often than not, a deliberate BIG LIE.

Soon, the troops were loaded, the mooring lines dropped. The Seacats' engines started, the ship slowly sailed across the harbor, Alcatraz Island passed by off starboard then, it was under the Golden Gate. Most of us replacement soldiers stood on deck watching The Golden Gate pass overhead. The Seacat picked up speed as it entered open ocean water.

It was a lovely day with a light breeze coming from the Pacific, the bright sun shown overhead. California gulls soared around the ships mast and superstructure. An occasional pelican dove headfirst into waves, trying to catch lunch.

It seemed only a short time when distant islands passed by on the starboard side of the ship. These were The Farallon Islands, a National Wildlife refuge for sea life.

Gulls might follow ships entirely across an ocean subsisting on bits of food thrown overboard after meals. Small fish, killed by The Seacat's spinning screws, were a continuous source of food for hovering bird life. Soaring gulls were a constant source of comfort, my friends and a connection to the world I knew.

As The Seacat entered the open Pacific, dolphins started to play chase around the bow of the ship. Hanging over the side for several hours, I watched the show dolphins were putting on. There were possibly a dozen of them playing chase beside the ships bow. Dolphins continually dove, swam under the ship and chased the bow from the other side. Possibly some continued to chase the ship entirely across the Pacific. This became a daily diversion from the boredom of endless days on shipboard, a pleasure for me to watch.

The increasing ocean swells were not immediately noticed. As they were getting bigger, the ship began gently rolling and pitching in the seas. For a time while watching dolphins, it was easy to forget reality. Much later, when leaving the rail it became evident what the pitching and rolling could do to a landlubber. For the first time in my life the nauseating experience of sea sickness swept me.

The Seacat did not have stabilizers, structures attached to the hull that minimized roll, pitch and yaw. Stabilizers serve as fins, working to stabilize ships, keeping them on a steady straight and level course in most seas. Ships so equipped, were much more comfortable for passengers. The Seacat plowed forward now at the mercy of wind and wave.

Soldiers were packed cheek to cheek both above and below decks. Most food served was little more than cold meat. Breakfast consisted of some make believe eggs (reconstituted powdered) fried on a community fry pan, cooked oatmeal cereal, a slice or two of spam and coffee, maybe. While on shipboard troops were given small portions of food. Probably this was to minimize the mess resulting from seasickness and there was plenty of that. This was real Spartan fare.

Mostly just to give bored soldiers something to do one would likely be given an assignment as interior guard. This was a nothing job. The only requirement was just man a post designed to keep soldiers, crew and dependents separated.

Sleeping quarters were very cramped. The hold, or large cavernous place below decks, contained steel supports for sleeping bunks. Five of them were stacked on top of one another.

When one was lying on his 'rack' or bunk there was only about eighteen inches from his face up to the next bunk. This left hardly enough room to roll over. Men were constantly nauseated as seasickness plagued soldiers during rough weather. With some, gently rolling seas were enough to keep them nauseated.

It was best to have a top bunk. Soldiers lying in a lower bunk would immediately know if a man above him was nauseated. Some lessons were learned very quickly. Within a few hours floors or decks were covered with vomit and were very slippery. Crews were assigned foe clean up detail, but they could hardly keep decks dry.

When we were at sea several days, when it was announced that we were headed for, Manila, sometimes referred to as The Pearl of the Orient. The Philippines were just a spot on the map to me at that time. Yes, I was aware a violent sea battle was fought at Leyte and General Macarthur's' famous wade ashore at The Lingayan Gulf, but little else. Get ready Grey, the Philippines will be home for the foreseeable future, better get acquainted. It was announced The Hawaiian Islands passed south of The Seacat's course, the previous day. We soon entered tropical waters.

Air circulation inside the hold was practically nonexistent. No fans circulated outside air into the interior of the ship. Hundreds of hot sweaty bodies produced gallons of sweat in tropical heat, raising humidity levels in the interior of the ship to near 100%. As we entered tropical waters, sleeping below decks was most uncomfortable, making it impossible to get sleep while in the hold, it was so hot and humid. Most soldiers decided it was far better to sleep on deck, even if it meant sleeping on solid steel. At least the air was clean, refreshing, the temperature and humidity, bearable.

There is a long standing rule in the military about never volunteering for anything. Don't believe it. A call went out for someone with meat cutting experience. This sounded better than drawing interior guard. I volunteered, reporting to the meat shop for duty immediately when the announcement was made. What a racket, excuse me, that turned out to be.

For about two hours each morning I put various chunks of meat on the band saw and cut them into steaks and chops. Blocks of frozen beef were cut into small pieces enabling the grinder to accept them to make hamburg. I also operated a meat slicer producing thin cold meat slices for the troops. (cold meat slices, usually bologna or other floor sweepings made into processed cold meat rolls, were referred to by the soldiers as 'horse cock'. Since this was very simple meat cutting it did not allow the use my skills as a meat cutter. It doesn't take a meat cutter to operate a slicing machine or a band saw.

My reward? I was privileged to eat in the crew mess and was excused from all other duty. These were first class meals, with fresh meat and vegetables. The morning we entered Manila Bay, I drank fresh milk from the cooler. All other soldiers were given reconstituted dried milk. That stuff tasted like chalk, not at all like fresh milk although it probably contained some nutrition. Nearly all soldiers refused to touch it, no matter what the nutrition might be.

There were new sea creatures to watch. Flying fish could be seen from the ships rail. They seemed to leap from the top of a wave and fly to another. Flying fish are about a foot to a foot and a half in length and have large pectoral and anal fins that enable them to glide and control their flight a considerable distance, perhaps flying as far as 100 feet. Some have made the observation flying fish actually flap their pectoral fins to remain airborne, making a controlled flight.

Flying fish eat plankton growing near waters surface. The purpose of their flight is to find food. What a remarkable adaptation.

It was also fascinating to watch sharks chase after various bits of garbage dropped from the fantail. (The fantail is the deck on the stern of a ship). They would "hit" a wooden box. When a shark 'hits' something it swims fast directly towards the subject starting from some distance away, opening the jaws wide immediately before biting or hitting the object. If, on contact they decide it indeed is food it is seized and torn apart. Sharks followed ships constantly, just as the gulls. Some of them seem to stay with the ship entirely across the Pacific.

When crossing the International Date Line or the Equator for the first time, by tradition, a formal ceremony is required for the uninitiated. All who cross must be properly ushered into King Neptune's kingdom.

Two misfits, (men who refused to learn how to live with others) were found to represent all of us who were crossing the Date Line for the first time. These soldiers were escorted to the deck and blindfolded. After much nonsense, that is, speeches, containing threats and mock beatings, the two blindfolded misfits were made to drink some "Green Dragons Piss, which was made by dissolving atabrine tablets in sea water.

This mixture was harmless, but the taste of atabrine and sea water mixture was worse than awful. Drinking it while blindfolded was enough to curdle any stomach with predictable results. Atabrine replaced quinine, a long standby drug, was used to relieve the symptoms of malaria during WWII. Atabrine, was bitter tasting and turned skin yellow within a few days when taken on a regular schedule.

After a considerable amount of harassment the two candidates were to 'walk the plank'. A plank about a foot wide was fastened to a hatch cover so it protruded over a large tank of water. The boys were forced to walk the plank and made to believe they were going to plunge into the South Pacific.

Instead they walked the plank and fell into a tank of water placed on the deck. While it was a nonsense show, the candidates were thoroughly frightened. During the harassment they were reminded several times, rather forcefully, sharks were swimming around the ship and they were hungry.

After this nonsense we were all, two thousand of us, given a certificate attesting we had entered King Neptune's' Kingdom. Somebody went to some effort, the certificates actually had each soldiers name hand written on it. Distribution following the initiation ceremony took considerable time. But it gave us something to do. Eventually everyone had in his possession his personal Certificate, proof he had entered King Neptune's Kingdom by crossing the International Date Line.

For the most part, the days aboard ship were monotonous, but there were notable moments of excitement. One day a floating mine, left over from the war, was sighted about 400 yards off the starboard bow. It was floating free, having come loose from its moorings. A loose ready to explode mine, presented an extremely deadly danger for shipping passing through the area. Floating mines might be carried by water currents to far distant waters remaining functional for decades. Any sighted, must be destroyed, before a passing ship accidentally hit it. The Seacat had been used to transport soldiers during Pacific battles still had a three inch rifle mounted on the fore deck.

The first attempt was made when sailors, part of the ships crew, took carbines and tried to hit a 'horn' on the floating mine. The Captain would not place the Seacat closer than a hundred yards from the mine for fear when it exploded it might damage his ship. That is not close enough to hit a small target like the horn on a floating mine with a carbine. The ship rolled and pitched even more violently when forward motion stopped. Unable to hit the mine with carbines, ships Captain decided to explode it with his deck gun.

The Seacat remained stopped, the gun, uncovered and a clip of ammunition inserted. The gun crew, while poorly trained, was successful in operating it rapid fire. They lay down a barrage of shots. Clip after clip of ammunition was fired with most shells missing the mine entirely. The deck gun roared, plumes of water rose high into the sky. As the crew grew increasingly frustrated, the military language around the gun mount turned the air blue.

Finally a lucky shot hit the mine which immediately exploded harmlessly in the ocean. The column of water produced by the blast was certainly impressive.

Manila, the Philippine Islands

After 28 days of relatively calm seas, the Seacat entered a channel leading between several of the Philippine Islands, including an erupting Bulusan volcano, it passed to starboard. Very early the following morning The Seacat entered Manila Bay. The sun was rising behind me, when Corrigidor and Bataan passed in easy view off the port side, surrounding green jungle growth glowed in the rising sun. The Seacat turned to starboard as it entered Manila Bay, the ship increased speed, possibly the Captain was eager to discharge his dirty, seasick passengers so he could have "his" boat cleaned. It might be a question whether he or the soldiers wanted the voyage to end as soon as possible. It is probably fair so say all soldiers wanted "the hell off that boat" ASAP.

While the Seacat was traveling across open seas the air temperatures did not seem oppressive. The morning we started in a channel between the Philippine Islands we were no longer receiving fresh winds blowing across open water. On deck that morning I was hit with the most oppressive temperatures and humidity imaginable. My thoughts were, "how is it possible to survive these conditions?. How long will this hell continue"? Will my good health remain, good health?.

Tropical temperatures smothering The Philippines were almost unbearable for this Yankee Boy. Thermometers constantly hovered around 100 degrees. In addition humidity remained around 98%. One always felt clammy and sticky, as if one's skin was always crawling.

Luzon is located only a few miles north of the equator where the sun's rays are almost directly overhead all times of the year. No 'seasons' mark the passing of months. Vegetation remains continuously green if not fresh and lush.

Monsoons last all year as rain showers fell most every day in mid-afternoon, regardless of what the calendar said. Possibly during the rainy season the showers might last longer than during the dry season. It was difficult to say what season it was as they were so similar on Luzon.

My body was not to be dry for the next several months as sweating is constant. Profuse sweating will cause the loss of electrolytes, making one feel weak and unable to remain physically active. One method to combat the loss of electrolytes from the body is to take salt tablets every day, this helps to keep one's body hydrated. It is true, with time, the body makes some adaptations to extreme conditions but this Yankee boy never could feel comfortable in the tropics.

The Seacat tied up alongside a floating dock, two thousand soldiers disembarked. Thankfully, at this point the military relieved us of all the Arctic gear we had been issued. The nice Lieutenant ordered us to drop that extra barracks bag in a warehouse, located dockside. It was written off each mans records, without checking whether it was all there or not. When we returned to the States, it was reissued for us to carry back to San Francisco.

Troops assigned to the Air Force were loaded onto waiting 2 1\2 ton trucks, commonly referred to as a duce and a half, and driven to Nichols Field, a replacement depot for the Air Force and the only airfield serving International flights to and from the Philippines.

Our quarters were pyramid tents arranged in very straight rows. A cute little concrete walkway lined along each side with small freshly painted white rocks leading to each tent. The walkway connected to the Mess Hall and Headquarters area. It was an attempt on someone's part to bring a bit of order and beauty to an otherwise stark area.

New arrivals were assigned to a tent and were told to wait for orders to report to an air base somewhere in the Philippines. In the meantime we were issued Class A Passes and were free to leave the base. The only requirement placed on us was to answer roll call each morning and respond to posted troop shipments.

Nichols Field, located a short distance from Manila City, made it easy for us to hitch a ride into the City on a military vehicle. It was customary for the drivers of military vehicles to give rides to hitchhiking GI's along the highway.

Now the real experience of my growing up began. Within a few days on Nichols Field I was exposed to much of the darker side of life. Within days this tender, in many ways a virgin, boy from Odessa, New York began an emotional change. The crude, brutal and primitive conditions existing in Manila were to have a profound lifelong effect.

When the first opportunity presented itself, a group of us hitched a ride into Manila. After the long sea experience and confinement on a ship it was good to travel on dry land. We couldn't get there fast enough. When first visited, Manila was a frightening scene for Manila lay in ruins.

There was not an intact building. Most were flattened or damaged beyond repair, including Government buildings. In some places not one stone had been left on top of another.

Pattern bombing is described as when dozens of bombers assemble wing tip to wing tip, nose to tail and fly over a target area releasing their bombs simultaneously. One can imagine the total destruction on the ground, caused by that action.

Manila was repeatedly bombed from the air and shelled by artillery. The Military claims, Manila was not pattern bombed, but it might as well have been judging from the amount of destruction observed. This, almost complete destruction was mostly the result of artillery bombardment. Scenes from Cologne, Dresden, Berlin or Tokyo or other pattern or fire bombed cities could not be worse.

After the conquest of Leyte and other southern islands, General MacArthur chose to invade Luzon from the Lingayan Gulf. After landing, thousands of GI's began fighting their way south passing Clark Field and on to Manila. Clark Field, bitterly contested by the Japs, is located about halfway from the invasion beach to Manila.

Driving Japs from that airfield was a top priority. When MacArthur seized Clark Field, he intended to trap and destroy Jap aircraft while on the ground.

It was concluded the Japanese army was short on gasoline as few aircraft were making attacks against the GI's either on the ground or in the air. When GI's finally occupied Clark Field, they found hundreds of Jap fighter aircraft (Zeros) and Zekes) had been left there. It didn't take long for G.I.s to destroy them all. The Japs would never have them available.

Manila also had been bitterly contested by the defeated Japs. Since General MacArthur had a special love for Manila he repeatedly asked the Jap General Yamishita to declare Manila an open city and not defend it. The Jap general likewise repeatedly refused. It was to be vigorously defended. Jap soldiers were eager to die for the Emperor; that is how they gained glory and won their way to heaven. GI's enthusiastically helped them on their way.

Vicious fighting continued as the Japs defended every piece of real estate they occupied. Of the thousands of Jap soldiers on Luzon at the time of our invasion, only a very small handful survived to become POW's. Fighting was reduced to combat between small patrols of soldiers, including hand to hand combat. After the fall of Manila smaller units of Japs continued fighting, hiding in the mountains of Luzon or on more remote islands.

Jap soldiers on these smaller islands were by-passed by the U.S. Army and Marines. After the surrender the responsibility for the "mopping up" hold out Japs was assigned to Filipino troops. Hold-out Jap soldiers continued fighting and killing several months after the formal surrender of Japan.

Some by-passed Jap units possessed fairly large supplies of arms and ammunition, making them dangerous for several months after the formal surrender.

While fighting their way south from the Lingayan Gulf, troops pounded the city with artillery around the clock before overrunning Jap positions, killing most Jap soldiers. When our troops entered Manila, fighting continued street by street, building by building. Jap soldiers were seldom taken prisoner or otherwise survived those battles. They fought to the death, believing surrender would bring dishonor to them and their families.

They repeatedly chose not to surrender. This remained true even when a small squad or a single Jap soldier was all that remained of larger fighting units. They continued the fight until the last man was killed or so badly wounded he no longer had a choice in the matter.

Fighting continued until every last pocket of resistance was taken by U.S. troops. Casualties among the Filipino civilians were also very high. As they were being defeated the Japanese murdered Filipino civilians, and POW's, of any nationality, as they were retreating. Many Filipino's took active roles, by seizing arms any place they could, either from fallen Japanese or American casualties and joining the fight beside American forces in defeating the hated Japs.

Sanitation in Manila was non existent. There were no sewer or water services, water was putrid. The atmosphere, filled with the stink of rotting garbage, urine and feces. These odors rose several thousand feet and could be detected while flying hundreds of feet above the ground. How could people exist in this stink hole?

We had won the war and supposedly it was finished. However, I was soon to learn, the war was far from finished. Huks were active insurgents on most Islands and Jap hold outs in remote places continued to prey on any Filipinos and or Americans they could destroy.

While in the tropics, we slept under mosquito netting hung from a metal frame placed around and over the cot. Netting was tucked completely around and under the mattress to keep all insects out. When arranged correctly this placed a person inside an insect free environment. Sleeping became possible. However when inside the netting, one must be careful not to allow bare skin to touch against the netting. Insects would bite through the very small mesh used to make mosquito netting. During the day, netting was placed on top of the frame out of the way.

Aerosol sprays called bug bombs, contained DDT an insect repellent, was used generously. Eliminating mosquito bites was essential to reduce the possibility of malaria infection. Soldiers found bug bombs had other uses. They could be punctured, allowing the contents to spray out continuously under pressure. When this was done the bomb became freezing cold.

Place a can of beer or soft drink in the freezing aerosol spray; it soon cooled, producing pleasant refreshment, so appreciated in tropical heat. Soldiers can be so inventive.

At first glance it seemed there was little positive about the city. As a country boy, relatively innocent, from a sheltered world I must learn to rise above my provincial background and look closely at places and events. One must ask why, how come, when and under what circumstances things happen, if real understanding was ever to come to me. Scenes were to come of apparent immoral behavior, cheating in any form and families selling their daughters sexual services and deception in the marketplace. Maturity was forced on me rapidly.

While Manila was shelled continuously for days, little shelling hit the old walled city along the bay shore. Inside the walled city, built by the Spanish many years ago, a swimming pool continued to exist undamaged. It was in use every day by hot sweaty GI's. Did it happen because its protection was deliberate? Did Commanders on both sides agree, "Blow Manila up if you must, but the swimming pool is off limits for destruction, DO NOT TOUCH IT".

Such a statement made by some high ranking officer sounds so probable. At least its fun to contemplate such an informal non spoken agreement might have occurred. One thing for sure, there was almost complete destruction of Manila, yet the ancient Spanish city was not severely damaged, interesting. There must be an untold story behind that.

At war's end, the Filipino people were demoralized and scarcely able to help themselves. I first visited Manila only a few weeks after the end of hostilities. Most of our fighting soldiers had been returned to the United States.

Relief Agencies had not yet appeared on the scene. Food, especially, was in critically short supply. Clothing was in tatters. People with serious physical disabilities such as cataracts, leprosy, loss of limbs, untreated wounds or severe burns, wandered aimlessly in the streets. Severely disabled or otherwise crippled people were constantly begging for food.

Nearly naked children could be seen stealing or pimping in the streets. Prostitutes were plentiful as families struggled to survive any way possible. People had open, running sores covering their bodies. With no sewage available, street gutters were used as latrines. During the first few months after the fighting ended, there were no relief measures taken. People were left on their own to survive the best way they could.

The United States was not long in sending food, medicine, clothing and skilled people to help the Filipinos recover from the war experience. Relief supplies were coming, but the need was great.

The Philippines were scheduled to be given their independence from the United States on July 4, 1946. They were to become an independent Nation.

With considerable empathy, I witnessed the overwhelming pain, trials and horrors experienced by innocent people and have since remained very much against anything that appeared as armed conflict. When hawks say war, inevitably people suffer endlessly. "Worth it," hawks claim. "Worth what to whom"? At what cost?, should be the immediate response. Man must find a better way to resolve conflicts.

Two beasts of burden were seemingly abundant in Manila. Slow moving water buffalo carried burdens of all kinds. When not working, they carried kids. Buffalo made fine pets for a family. The other animals were horses, a welcome sight. Horses, any horse looked and smelled great since was a bit lonesome for my own, far away at home.

Horses in the Philippines were all introduced at one time or another by various western cultures. It is known the Dutch introduced a few during the early 19th century. When Spain controlled the Philippines they brought with them their finely bred Andalusian horses. Apparently most of these animals died out or were worked to death or maybe the Japs ate them as none remained at the close of WWII.

Horses I observed on Luzon were quite small, much too small to be ridden by people, weighing only around 300-500 pounds. However, they did seem to retain some of the physical characteristics of the larger Andalusian originally introduced by the Spanish. At this point it's difficult to understand what actually happened with horse breeding in the Philippines over the past three hundred years.

They might have originally been Dutch or Spanish stock or perhaps the horses I observed on Luzon were actually the Mongolian pony, a small animal, originally imported from China. Perhaps through selective breeding they could have adapted to tropical heat. These observations are those made by a very young man with limited experience determining breeds of horses.

These small horses were put in a light harness and used to pull a light weight two wheeled cart carrying no more than two people or perhaps they might carry smaller loads on their back. Horses appeared plentiful, one could conclude, quite inexpensive on the market. Likely they were relatively inexpensive because of their limited value in the work world and they required considerably more care than water buffalo (caribou)

On the other hand, water buffalo certainly were well adapted to tropical environments and the work they were required to perform. Water buffalo, sometimes referred to as a caribou, were found in the more rural farming areas were they were the choice beast of burden. They were mild of spirit, easily trained and responsive to the commands, even those given by children. They might be used to carry most anything, including bundled bamboo, bags of rice, or kids. In rural areas every family seemed to have one.

Basic transportation showed the ingenuity of the Philippine people. Surplus, abandoned or reconstructed Jeeps or other discarded military vehicles were recovered, repaired and put into running condition by the Filipinos.

Jeep bodies were salvaged, entirely reconfigured and rebuilt to make a taxicab, referred to as Jitneys by G.I.s, capable of carrying 6-8 people in comfort.

These vehicles were decorated with bright cloth. Mounted on an overhead frame, a tarp or sheet of nylon provided shade to passengers. Tassels, flags, ribbons or any other flashy material might also be attached to attract attention.

Where drivers found them are beyond imagination, they mounted loud cowboy type musical horns on their Jeeps. These vehicles made quite a show as they whipped around the streets of Manila.

Larger abandoned trucks for instance like weapons carriers or a duce and a half (2½ ton) were torn down leaving only the engine, transmission and differential, attached to the frame. The vehicle was completely reconfigured into a bus. When finished, the owner was in business.

The Philippine government did not exist prior to 1946. Government licensing or regulation of motor vehicles had not been developed. The United States was pulling out, leaving government to the Filipinos themselves. Vehicle inspections, registration or operators licenses were not required. Whatever people could imagine and bring into existence, became a means of transportation. Safety equipment, forget it. Homemade buses filled the highways around Luzon. They could be seen on every highway carrying a load of passengers with possessions piled behind a fence like frame on the roof and they liked to drive fast, very fast.

These transformations did not occur inside an air conditioned repair shop with an abundance of power equipment available. Filipinos, Flips, made these accomplishments in their back yard with a minimum of tools. Cuts in metal were made by hand, since little electric power existed. How was metal welded? Was a charcoal fire used with bellows to coax a charcoal fire to reach high enough temperatures necessary to melt and weld steel?

Open markets were set up and operating in nearly every village on Luzon. Shops might be erected from broken boards salvaged from some wreckage. Shells of remaining buildings had been cleaned out. With debris removed it was possible to display merchandise for sale.

For the first time in my life I saw meat hanging in open air from a hook over a table, possibly all that remained from some animal that died, who knows how. Death could be the result of either a hit on the head, disease or starvation, any dead animal was utilized in some way. Clouds of flies swarmed on all bits of meat or blood lying on the table where cutting was done, no refrigeration was available.

From a quick glance, the animal the meat came from originally was impossible to determine. It's a wonder disease and digestive problems were not more prevalent than they were.

People pushed and jostled one another in the markets when trying to get close enough to vendors to make a purchase. It was the custom in the Philippines society to bargain price with vendors for most anything offered for sale. Under shortage conditions, few negotiations occurred.

Customers paid whatever the vendor asked and were glad to have the opportunity to feed the family.

Much of the meat appeared to be red meat that is came from cattle, water buffalo, sheep or horses. Chickens might make a brief appearance however they quickly disappeared, sold live before slaughter. The vendor would kill, pick and clean at the time of sale, if asked to, however most chickens left the market place alive.

When they were to be eaten, the complete bird was destined for consumption, including the entrails. Chickens tended to be cheaper because they were quicker to raise and they needed little care, finding most of their own food picking seeds from the ground. Also they could be moved from home to market quicker. A basket of chickens could easily be placed on top of a bus for a quick daily trip to a local market. This resulted in a more abundant and fresher product and meant quick cash for the producing peasant.

There might be fish or other seafood offered. A limited amount of fish, netted in Manila Bay, might be offered for sale along with fish caught or netted from rice paddies. Women and children were frequently observed fishing with a bamboo pole and line in rice paddies.

There were other beasties offered for sale. I had no idea what these things were as they were far beyond my experience to identify. However, I harbored no urge to taste test. With the high constant demand and small supply most food, most merchandise was sold out by mid morning.

Possibly one of the many health problems most Filipinos had was with their diet. The diet of many Filipinos especially for those living on remote primitive islands was often inadequate to maintain good health. Many diets were deficient in protein causing generally poor health. In addition, their bodies cannot successfully overcome infections as the immune system is weak. This makes their bodies vulnerable to trauma of any kind.

Generally, Filipinos were either, farmers, craftsmen and/or merchants. With abundant tropical woods available, highly skilled craftsmen developed a market for furniture, kitchen utensils and aesthetic pieces of carved wood.

There was no shortage of beautifully carved trays, knife, machete handles or religious symbols. In a market place or from roadside stands, intricately carved pieces were offered for sale. Wood to make them must have been brought to central Luzon from some distance away perhaps from other islands as few trees seemed to be available in Central Luzon. Mahogany was abundant as was another beautifully grained soft wood, I could not identify, suitable for carving.

The bright spot on the local economy, there were plenty of U.S. soldiers and sailors present. Privates or grade 7 enlisted men were paid $50.00 per month. This was a fair sum at the time but never quite covered all the activities of the average service person wanted to do. After all, there was an abundance of alcohol and women available for a price. Filipinos thought $50.00 was a grand sum of money and they wanted some of it.

Since many Flips possessed merchandising skills, merchants found items to sell to the 'well heeled, rich' Americans.

Much discarded military hardware from both the U.S. and Jap armies was just lying around the countryside. Metal, especially aluminum, recovered from fallen aircraft, was prized by the Filipinos, it might be cut and fashioned forming salable items, like knives, machetes and jewelry. Plexiglas, used in forming the windows and blisters on aircraft were cut into intricate pieces to make handles for utinsels and knives. Filipino artisans made most colorful items from salvaged materials.

Filipinos possessing artistic talent, set up shops in the marketplace. They might have in their possession an easel, a few paints or they might use other materials adapted for use as paint.

For a price, seemingly inexpensive to a G I, his portrait could be painted immediately on the spot. Several paintings I observed looked very good to me. The likeness produced seemed to be true and the colors real. Talented people could support a family in this way. This sure beats forcing a daughter into prostitution.

Those who possessed seamstress skills collected nylon parachutes making them into elaborately decorated clothing. Parachutes came in a variety of colors and textures. Chutes used by airman to bail out of disabled aircraft were of the highest quality. White nylon used in these chutes had a smooth glossy finish.

During combat operations, hundreds, perhaps thousands of cargo chutes were used to supply fast moving troops pushing against the Japs. In many situations continuous rapid re-supply by truck was not feasible. Dropping food, ammunition and medical supplies, items most needed by troops in combat, from aircraft proved a faster, more practical and successful method to keep troops on the attack.

A skilled seamstress could turn a chute into a formal gown, wedding dress, tablecloth, draperies or anything a customer desired. G.I.s ordered special made items to be sent home to a wife, sweetheart or mother.

The Chinese were especially adept merchants. Somehow they could find a way to do business under the most adverse conditions. They had regular, unending supplies of textiles, wood carvings, clothing, ivory and war relics such as knives, jungle equipment or Jap rifle could be found. Chinese markets could even supply a buyer with a highly valued Jap Officers sword, if the price were right. They seemed to be the first to recognize the market potential in a GI. Obviously they catered to the Army and Navy personnel stationed on Luzon where the most G I s were stationed.

The 13th Air Force occupied a large base at Clark Field, 60 miles north of Manila. In addition to docks at Manila where cargo ships docked, the Navy had a large base at Subic Bay, located a few miles west of Manila.

When the first of the month arrived with pay day for all military personnel, merchants did a brisk business. When the Navy docked large ships in Manila Bay, several thousand of sailors might be granted leave.

Soldier stationed on Nichols Field never left the base alone. They always traveled in groups of three or more. There was good reason for this practice. One would think the Flips would show gratitude to American soldiers for freeing them from the hated Jap domination, but it did not work out that way in practice.

When we left the base it was always in a group of at least five. Some of the group might want to drink, others wanted a woman. Those of us not drinking or seeking a female served to watch for any kind of trouble, hold the wallets, watches, rings or other valuables of those who became inebriated while in a bar or were busy with a prostitute.

Flip criminals lay in wait for a lone GI outside the gates of Nichols Field. Their purpose? To assault him, steal his money, watch or any other jewelry or valuables he might have with him.

Law enforcement was poorly organized with government control, in fact, virtually non existent. No Philippine police were present on the streets, only the MP's who watched over military personnel. As one might expect, crime was everywhere. A GI traveling alone was a target, sure to be attacked, beaten or rolled. Thugs preyed not only on GI's but their own people. None were safe from their brutal activities.

Prostitution was everywhere. In Manila, available girls numbered in the thousands. Many tricks were turned inside a girl's home with a member of the family, usually a brother pimping for her.

There were also large, organized Houses of Prostitution that seemed and well run. A soldier or sailor could present himself at the door, pay his fee and take his pick of whatever girl was available at the time.

There were always strong men available to protect the girls or rob the inebriated customer. It was foolhardy for any G.I. to enter a House of Prostitution alone, whether inebriated or not.

Now was to come the time for Grey's awakening. One night two of my group, there were five of us that night, encountered a little kid, probably around 6-8 years of age, on the street who was pimping for his sisters. A deal was made; the five of us went to the girl's house where payment was made to her father. Two girls, presumably daughters, took two soldiers upstairs for an hour of sexual activity. The remaining three of us sat at the table making small talk with the girls parents. The process took only about a half hour when our buddies returned.

One remaining daughter, who was actually quite attractive, lay on the table in front of the three of us, wearing only a very thin dress that revealed every dip and curve of her young body while covering very little of it.

She put on a show, doing everything possible to entice any of the three of us to pay her father and take her upstairs. She was showing off everything with all the cute moves that normally entice men. This entire scene was sickening to me. I could not understand what or why this was allowed to happen.

My personal reaction to this was very prudish. How could people in a strong Catholic country like The Philippines be so immoral as to sell their daughters in their own home?

Not everyone possessed artistic talents. Prostitution, most any female could perform for money. Yes, they would sell their daughter's services to a G.I. for a price. The girls appeared to freely offer themselves, without protest, in their willingness to help support her family.

It seemed so immoral and evil. I wondered how my companions could possibly face their wives or sweethearts when they eventually returned home after such encounters.

Actually, it took several years for these lessons to sink in. I finally come to realize, this was necessary for survival. There were no jobs, no money, no social services, no food and little else. Where was the money? Obviously, only the American GI had cash.

Flips regarded the average G.I.s pay as a small fortune. They would do whatever was necessary to acquire that money, seizing any opportunity to survive. Methods to get cash included prostitution, robbery, theft assaults, fraud or trickery.

Females, young or old, did anything necessary to sustain life and keep the family together. If one of the daughters became pregnant, the child would be raised by her family as they accepted this responsibility. This was not true for the general population as Amer-Asian children were not readily accepted into the Philippine society.

This was also true of Jap-Philippine children usually born after the rape of a Philippine girl. Jap soldiers were not inclined to pay a prostitute. They simply captured one and gang raped the girl, usually murdering her after the attack. If she survived such assaults she was almost certain to have been impregnated. Children resulting from all soldiers, Jap or American, were to have a very difficult life.

Usually the group of soldiers I associated with went to smaller bars, one which might be the front room of someone's home. When GI's entered and sat down, young girls would immediately sit at the table, one for each soldier. These girls encouraged drinking, although they did not drink alcohol themselves, and were available for sexual services. It was seldom one of us became drunk but some utilized the sexual services offered by the girls. Mostly we just socialized in a friendly way and were not careless with money.

It was the week of my 21st birthday. In many ways, Grey was still a virgin, at least where alcohol was concerned, I had never previously been drunk, actually had never tasted hard liquor. Perhaps a little wine or hard apple cider, but that was all.

My buddies thought it was about time for me to grow up and experience another part of life. After some teasing, they convinced me to give it a try. First, some unbreakable rules must be accepted by all my buddies.

Since I had no idea how my behavior might be once I became drunk, I made it clear. Do not let me touch a "gook girl". (The term 'gook' was applied by G.I.s' to all Asians) I want to return to my wife at home clean and faithful. My friends agreed they would not let me indulge in any sexual activity no matter what.

We hitched a ride on a weapons carrier (a weapons carrier was about the size of a 3/4 ton pick up truck) and were soon traveling the few miles from Nichols Field to Manila. We found a small bar on the edge of the city went inside and took a table large enough for the five of us.

Grey started to drink straight whiskey. Since my buddies were treating me, the supply was endless. As usual, when we G I s sat down, girls appeared to sit with us. As alcohol took its effect, I started to get verbally abusive. At this point in my military career I had learned an abundance of four letter words and was using all of them to describe the military situation etc. In fact I became so obnoxious the girls left the table.

When I was so drunk I could hardly stand or walk, my friends decided enough was enough. We all caught a ride back to Nichols Field on another weapons carrier and were dropped off outside the gate. It was now required of me to walk past the sentries guarding the gate while supported between two of my buddies. The MP watching me stood in the roadway hands on hips watching every move as I struggled to stay upright and walk past him.

There was a regulation. All soldiers returning to base drunk must report to the "pro station" located immediately inside the gate for disinfection from any sexual conduct.

My guardians started convincing the MP s there had been no sexual contact. Success, the MP did not insist we all visit the pro station. Friends walked me a mile or so from the gate to our tent. Along the way I continued a tirade against the military, saying at one point "I would like to meet a 'chicken' officer and beat him up", someone was walking towards us. He responded 'here's one.' My buddies walked me out of his presence as fast as possible.

The next morning there was the expected headache. My buddies had placed me into bed, tucked mosquito netting around the mattress and left me to sleep it off. While coming out of the fog the following morning, a horse nickered. For a few moments, it seemed it was my favorite saddle horse, King, at home calling for breakfast. No such luck.

Another soldier had also been drinking himself silly. While in this condition he somehow gained possession of a horse and smuggled it onto the base. It was tied to his tent. Here was another cowboy far from home. It was a lesson for me. Grey has never again been drunk, been there done that. Since that night it has only been one occasional drink of wine, maybe to celebrate a birthday Christmas or some other family occasion.

Some officers were concerned about the welfare and morale of soldiers. A kind hearted Captain arranged for four trucks to take a group of us to a quiet bay in eastern Luzon so we could have a day at the beach. It was to be great fun, combining our efforts we gathered enough food to enjoy a picnic together.

When we arrived at the idyllic sandy beach along the southeastern coast of Luzon, we stripped off our clothing and made a running head first belly flopper into the gentle surf.

Being an accomplished swimmer, comfortable in deep water, it seemed OK to swim out from shore a few yards into deeper water, practice a few strokes and make a few surface dives. It felt good to be in warm water enjoying gentle physical activity. Pleasant, comfortable water flowing along my naked body felt like a little bit of heaven. I knew or cared nothing about the beasties swimming alongside me.

After several minutes swimming all other guys had left the water and were gathered huddled on the beach, Grey was the only one remaining in the water. Several had ugly red rashes rising across the skin covering torsos, arms and legs caused, by floating jellyfish. They were milling around, making what I thought was a big deal out of very little.

Being of little patience and new (a tenderfoot) to the tropics and the ocean it just didn't seem possible those jelly fish could bother ME. I took another running header into the surf. This time alone, all others remained on the beach, stung or not.

Floating jelly fish have long tentacles that trail in water ten or fifteen feet. Hundreds of stingers grow along these tentacles. When a human swims into one of these, the tentacles wrap around a swimming body penetrate skin, and inject a toxin, causing an immediate painful red rash. In a few moments Grey was stung across the chest and left arm.

The return trip to Nichols Field included a stop at the military hospital where several of my companions were left to receive an antidote for the poison in their bodies. This was another lesson for me. While I did not get an allergic reaction to the poison it remained painful for several days. Grey, when you do not understand the hazards of this environment, don't be so cavalier.

One night it became painfully obvious the shooting war had not ceased in the Philippines. In the middle of the night, a barrage of gunfire erupted around the perimeter of Nichols Field, about a mile or so from the barracks where troops were sleeping. An unlucky soldier, sleeping in a barracks jumped to full alert, awakened by the gunfire. While standing beside his cot, preparing to go to the latrine a bullet fired from a mile away came through the swale wall, hitting him dead center in his chest.

The boy was dead when he hit the floor. While sleeping in a barracks with 60 other soldiers, this bullet flew through the side of the building, passing above other sleeping soldiers, hitting the only one standing erect. No, the shooting war had not yet ended. This demonstrated just how dangerous this place could be, even on a supposedly safe military base with thousands of GI's present.

The United States had agreed to give the Philippines their independence July 4, 1946. The country was engaged in a power struggle. Philippine forces representing the government the United States supported were trying to eliminate Communist inspired guerrillas called Huks. Huks were backed by Communist Chinese who were attempting to hijack the newly formed Democratic government of the Philippines.

Huks were attacking Filipino Army and U.S. military installations. These were small affairs, with about a dozen Huks attempting to raid military installations. Their purpose, to steal whatever war materials they could carry away. If any G I interfered with their mission, firefights were instant and intense.

Huks were known to creep up and hack any soldier in sight with a machete. Several times, G I's including those not on sentry duty, were badly wounded or killed during these encounters.

While at the Replacement Depot on Nichols Field interior guard duty was frequently assigned to protect the tent area where soldiers slept. Several of us served a tour while waiting a more permanent assignment.

One night my assignment was to patrol the pathway winding around tents and confront unwelcome visitors. Sentries were issued a .30 caliber carbine with loaded 15 shot clip.

The night was black with almost no light making it difficult, almost impossible, to follow the path winding between tents. It was too dark to actually see the walkway. It could only be followed by the feel of gravel under one's boots. The assignment, walk a four hour tour of interior guard duty from midnight to 4 am. (2400 hours to 0400 hours military time). Alone, in a black night, the potential existed for some Huk to make an easy kill, me.

For the first time in my life I knew real fear, yes terror. The best defense, make no sound, move slowly, watch and listen carefully, keep that carbine at the ready position, loaded, cocked, safe off and ready to shoot. A Huk with a machete could easily ambush a sentry in this situation. It was so dark one could walk very close, within arms length, of a standing person and fail to see danger. While armed with a .30 caliber carbine, fear made effective defense seem impossible. Personally, my fear was more of what that machete could do than a bullet.

I don't mind confessing to you, the reader, of my most unsettling military experience. When one is frightened and feels there is no defense from receiving extreme pain and/or that death could be inflicted, the mind fantasizes on a myriad of possible escapes.

While in this situation, self-mutilation passed through my mind, like shooting myself in the foot and claiming it was an accident. Other imagined scenarios were desertion, or bribing another soldier to take my place or finding a place to hide for the four hours also seemed attractive escapes. If I could have convinced myself to try, I would have started swimming home.

These, and more, of fantasies raced through my mind, anything to escape that fearful situation. While in the midst of great fear God, my Father's voice, could be heard. Grey, do not let fear destroy your life. Conquer them. With all these fears, self control prevailed, four hours passed without incident. I had successfully conquered disabling fear. That night served as a powerful lesson about keeping control of ones emotions while struggling alone in a potentially deadly dangerous situation.

Another lesson learned: Never judge the actions or decisions made by anyone who has been placed in such an emotionally stressful situation. Monday morning quarterbacking is nonsense. Politicians, law enforcement personnel, judges and people in general should be tolerant and more understanding of reactions to extreme emotional situations. Stop pointing fingers of blame. One cannot possibly understand the power of such emotions until one has personally experienced them.

My mind could not dismiss the frightful image of a gleaming machete swinging towards my head. Although it had been more than 63 years since that night, haunting images remain, playing out over and over again inside my head at unexpected moments.

What is the lesson to be learned from this experience? While none of the wild thoughts were strong enough for me to act on, there existed a strong instinct to survive without injury. Credit a strong family influence here. Explaining an act to my loved ones that could be called cowardice would have been impossibly embarrassing. No one in my family would believe a story about an accidental firearms discharge. They all expected more of me.

All of us need to develop more patience and understanding when people are so overwhelmed by fear they act out a fantasy. A rush to judgment is not helpful, needlessly destroying human lives.

Remember, in the case of soldiers. Our Society placed young men and women in harms way by sending them to fight a war. Society itself must be held responsible for any and all consequences.

Prison sentences and Bad Conduct or Dishonorable discharges that deny social acceptance and/or Veterans Administration benefits are not solutions to a soldier's situation. Research, providing continuous further understanding and adequate funding for a soldiers care through The Veterans Administration serve a soldier more positively.

The lesson of self preservation was to become extremely important months later on Clark Field when it again was my turn to be an interior guard.

A C54 Skymaster was approaching Nichols Field, coming in for a landing at the end of a long flight from Hawaii. The aircraft was reportedly carrying medical supplies needed in the military hospitals at Manila and Clark field,

Several replacement soldiers with specialized skills and a BIG shipment of anticipated MAIL to be distributed to soldiers stationed in the Islands were reported to be on board.

It was a beautiful sight, the great four engine aircraft gliding gently onto the runway, the sound of synchronized slowing engines. It looked perfect, how could anything go wrong? The wheels touched the runway with most of the load balanced on the main gear. The aircraft slowed, weight slowly shifted forward to the nose wheel, as the wheel gradually bore more of the aircrafts weight, it suddenly collapsed. The C54 nosed down with the fuselage grinding along the runway generating considerable heat as the belly of the fuselage became red hot. Gasoline and gasoline vapors remaining in the tanks caught fire and exploded. The aircraft slid to a stop in front of the terminal, sat smoking a few seconds then exploded in a fireball. All on board were lost, no mail or supplies salvageable.

Finally my name was posted to an assignment. It was to the 2nd Air sea rescue Squadron on Clark Field the day the order was posted a duce and a half was ready to transport several of us.

From Manila to Clark Field is a distance of 60 miles. Abundant scars of war dotted the countryside. Bridges were down. Temporary floating bridges had been placed so vehicular traffic could cross wetlands. Highways had been severely damaged by shell fire. Few craters had been filled enabling traffic to cross. In many situations traffic was forced from the highway to avoid deep unfilled craters.

To successfully skirt unfilled craters, drivers needed considerable skill as roadways were often built up or elevated from the adjoining rice paddies. An error in judgment could cause a vehicle to roll from the highway into a rice paddy landing on its side or worse, upside down.

During this trip, lasting several hours, passing through Central Luzon, I made a few observations about life in the Philippines. Let me digress here a few moments to describe Filipino homes, livestock and how poor Filipino's made a living. This was a new social and economic environment for me, situations never previously understood.

Most buildings were destroyed or damaged from flying shells or flame throwers. However native residents had rebuilt their bamboo and swale huts so in that respect were living comfortably. In rural areas bamboo, a plentiful grass, is readily available and can be used to quickly erect living quarters.

When making a mat of swale, lengths of bamboo are cut to length then split into long narrow strips with a machete (a machete is an all around general purpose knife, about a foot and a half long, used in many tropical areas around the world).

Green bamboo is very supple, when split lengthwise, can be woven to form a flat mat of any size, depending on the purpose of its intended use. Swale mats might be used for such diverse purposes as to make building walls, a sleeping mat a floor or even a pigpen. Swale could give privacy, kept the monsoons out of living areas. Also, it made excellent fencing to hold livestock.

While it doesn't look like it, bamboo is a grass that grows tall and abundantly. It might tower as much as eight to ten feet, growing naturally, in abundance, on central Luzon, where temperatures are always warm and rain occurs almost daily. These two environmental conditions are needed for bamboo to flourish.

Multiple mats can be stacked together making mats thicker. Generally bamboo mats, now called swale, were lashed to a pole framework where they formed the sides of a more permanent building that might be occupied or used for storage of tools.

Filipino homes on Luzon were simple. Four, six or eight bamboo poles, about ten feet long were planted firmly in the ground forming in a rectangle.

Floors were bamboo poles lashed tight against one another to form a flat surface making a floor, placed about 6 feet above the ground. Floors were fastened into place with vines or perhaps nails if they could be found.

Swale mats were tied over a bamboo pole framework. Ladders were placed to give access to the elevated floor, where sleeping quarters were placed.

Swale mats, the size of an intended floor were covered with a paste of some sort spread on with a brush, another mat was placed over the top of the one covered with the mud mixture.

The sandwich was allowed to dry when the mixture, made of some easily found mud, became like concrete, binding the two mats of swale together. When finished, a durable solid floor had been made.

Sleeping quarters were placed on the elevated floor, where a family could keep dry should flooding occur during heavy rainstorms. The rainy season turns seemingly solid earth into mud, not a comfortable place to sleep. Most daily living however, was conducted on the ground level. This included cooking, dining and socializing.

A building made of swale had plenty of air circulation throughout, a condition required to keep the inside of any building comfortable for people. Roofs were made of overlaid palm fronds or corrugated tin. Palm fronds successfully kept most rain from coming inside a building

Land immediately to the north of Manila on the island of Luzon was mostly wetlands filled with rice paddies. Two volcanoes could be seen in the distance both east and west of the lowlands.

A system of dike's and canals made possible the control of water into each families land. Filipino poor worked the land planting, cultivating and harvesting rice plants. Seemingly everyone in a family performed the stoop labor necessary to plant, cultivate and harvest rice.

Older women or young children frequently caught abundant small fish in rice paddies using the traditional Huck Finn method that is a long bamboo pole with hook and line attached.

In addition to water buffalo, other livestock might include pigs and chickens. Animals seemed to free roam. Chickens wandered around the ground level of the home and adjoining areas. Fences for hogs or other small livestock such as sheep or goats, were made of bamboo poles. Wire fencing being too expensive or not available.

Outside the city, water buffalo seemed to be the principal beast of burden. Very few, if any, horses were kept outside the city. Horses were probably too expensive to keep they carried small burdens and were helpless in deep mud. Larger, stronger water buffalo were hitched to farm implements, like plows or wagons. A travois similar to those of the American Indians might be used for some burdens.

Water buffalo are large, powerful animals, able to handle heavy burdens. They are also docile enough that they can be controlled by children, who make playmates of them.

I was riding in a convoy of five 2½ ton trucks (usually referred to as a duce & half truck) to Clark Field along the only highway leading north from Manila towards The Lingayan Gulf. This was the landing point on Luzon where General MacArthur took that famous walk through the surf making the photograph widely used to announce MacArthur had returned to the Philippines and was closing in on Manila.

About 10 miles south of the village of Angeles, north of Manila, the convoy was stopped by US Army MP's to avoid a battle being fought across the road ahead. This was a shoot out between Communist Huks and the Philippine army. This firefight went on for several hours with shooting coming from both sides of the road. It did seem to be mostly small arms but grenades were also being used. Absent were the sounds of heavy explosions coming from larger guns and shells.

After several hours of continual fighting the convoy was allowed to continue its travel to Clark Field. Causalities on both sides are expected during these shooting encounters. Our trucks passed across the battle zone.

Yes, there were bodies, clad in both military and civilian clothing, lying in rice paddies. It was plain to see why some observers of battle scenes describe the carnage with phrases like "it was a blood bath or waters ran red.

Blood was everywhere as were torn bodies lying, mutilated from grenade and gunfire. Emotions overwhelmed many of us as riding on the truck that day, no one had ever witnessed torn dead bodies, lying in blood. It was such a shame. Flips killing each other over issues no fighter could ever understand.

In the center of the Village of Angeles the convoy turned from the north south highway to the west towards Clark Field. Angeles was a filthy village. If possible, it was in worse condition than Manila. Sewage and garbage odors filled the air.

There were a few positives however. Shops lined the streets and in the central Plaza items were offered for sale to G.I.s. For these merchants, business was very good. Soldiers have money and are very willing to spend it. Angeles will rise out of its ashes. It is important to remember, the area around Clark Field was hotly contested by the Japs. The fighting here was intense and prolonged as both The U.S. and Japanese army wanted control of this airbase, the largest in the South Pacific.

Civilian populations suffered very heavy casualties. The destruction of their homes and infrastructure was nearly complete with highways badly damaged, civic buildings destroyed, homes burned, bridges knocked down and irrigation systems rendered useless.

2ND AIR SEA RESCUE SQUADRON

The convoy finally arrived at Clark Field, my assignment, the mess hall located in the 2nd Air Sea Rescue Squadron.

Clark was the largest air base in that part of the Pacific. After the Japs were driven from The Philippines Clark Field became a major base for B29's to bomb Iwo Jima and Japan. The runway was long by the standards at the time being exactly 5,000 feet. However, this was scarcely enough runway for a loaded B29 to attain sufficient flying speed to become airborne.

Watching a B29 "takeoff" was a different experience. The pilot revved the engines to top speed, rolled down the runway holding the aircraft on the ground while continually gaining speed. He did not pull back on the yoke to become airborne until he had used all 5,000 feet of runway. Once in the air he did not head for the furtherest star, he remained slightly above treetop level for several miles while the aircraft continued gaining airspeed. When airspeed was attained, he took it upstairs. Watching one "take off" the B29 was over the outskirts of Manila, the aircraft was still only 200 feet or so above the ground. There was no bomb load, only an empty aircraft.

The 2nd Air Sea Rescue Squadron was organized in 1943. The Squadron was assigned to The South Pacific, moving several times as our forces marched across the Pacific. They rescued hundreds perhaps thousands of fliers shot down by enemy gunfire.

In September 1945, at Wars end the Squadron was moved to Clark Field on the Philippine Island of Luzon. Major James "Jungle Jim" Jarnigan was placed in Command.

Rescue Squadron's enjoyed the unique distinction of being one of the few military combat units whose work consists of saving life rather than destroying it.

Usually one understands the term' mission' is used to describe a single assignment. In the 2nd Air Sea Rescue a <u>MISSION</u> was called a **<u>SNAFU</u>**

(An Acronym for <u>S</u>ituation <u>N</u>ormal <u>A</u>ll <u>F</u>ouled <u>U</u>p)

Outside the entrance an attractive sign designating the 2nd Air Sea Rescue Squadron stood another smaller sign standing proudly and just as attractive, reading SNAFU SNATCHERS, a silent testimony to the purpose of our missions.

The 2nd Air Sea Rescue Squadron was not very large, consisting of about 250-300 Officers and enlisted men. The flight line was located about a mile from the Squadrons assigned living area. We had, assigned for our use, 12 aircraft, four were B 17s and the other eight were PBYs also known as the Catalina (type OA-10A) most commonly referred to as The Cat.

PBY CATALINA or CAT

It seemed to me there was always a high level of morale in his Squadron, its Officer's, nothing as I have described previously. These were honest people united in the work of saving lives. Daily our lives depended on each others integrity.

During WWII Catalina's were mostly used as patrol aircraft. Cats saw extensive service in Europe where the British and Canadians used it to hunt Nazi submarines. The U.S. Army loaned several of them to Brazil also to meet the submarine threat. Cats had a range of about 2,500 miles, with twin 1200 horsepower engines and a wing span of 104 feet.

The PBY Catalina, also lovingly referred to by men of The Snafu Snatchers Squadron as The CAT. CATS were flying boats called amphibians with a tricycle landing gear making possible safe three point landings on either runway or water. The Cat's fuselage was shaped like the hull of a boat, making water and runway landings equally possible.

Mounted on the wing above the cabin were two 1,200 horse power Pratt and Whitney engines." Cats" were slow with a cruising speed of 117 mph and a maximum speed of 179 mph. It had a flying range of 2,545 miles. When flying a search, it could fly as slow as 90 mph and remain airborne. These reduced speeds were essential when searching for survivors, whether over land or sea. When flying a search it is most desirable for aircraft to fly low, slow and remain aloft for long periods of time to achieve maximum effectiveness when seeking to find wreckage or personnel whether on land or sea.

On the tips of the wings were mounted pontoons, easily raised up to wing level for flight or lowered to water level when making water landings. Wing tip pontoons gave the pilot more stability when making water landings.

Wing tip pontoons could prevent a wing from catching the top of a wave. This might cause the aircraft to spin around, to its destruction. While in flight, pontoons were raised to wing level, appearing to form an extension of the wing.

On the port side of The Cat in the fuselage under the wing, near amidships, a door or hatch opened for loading and unloading. When parked on dry land, a ladder was needed to climb aboard, the opening being about five or six feet above ground level. When sitting in water, the hatch was only two or three feet from water to hatch opening. Providing the water was calm, this made it easy to load and unload people or cargo, when parked on waters surface.

It was generally, through this hatch, crewmen launched yellow rubber rafts with rescue personnel to affect the rescue of crash victims. On the other hand, if the water was rough with waves four feet or more, the hatch door could not be opened, forcing crewmen to launch rafts from the blisters, a much higher position from the water.

Collapsed and folded rubber rafts were placed in the main cabin. When necessary to launch, rafts were removed from the storage locker, held outside the cabin and instantly inflated with a CO_2 bottle.

Rafts became the means necessary to either go ashore or to paddle close to a survivor for a pick up. When the raft returned to the aircraft it was easily deflated, refolded and stored away before take off.

Rubber life rafts, regardless where initially placed, (on aircraft or ships) were always colored yellow on one side, sky blue on the other. If someone <u>wanted</u> to be sighted from the air, he turned the raft yellow side up. If on the other hand, the survivor did <u>not</u> want to be seen by an enemy searching for him, rafts were turned blue side up making them virtually invisible from the air.

The Cat was used extensively in the South Pacific for making rescues of downed airmen. Cats were scheduled to follow fighters and bombers to the target area, orbiting along the assigned return route or near the strike zone. From these positions a Cat would be close by for rescue when any of our aircraft, hit by enemy gunfire, found it necessary to "ditch."

Should any of our attack aircraft making a strike against the enemy, broadcast a May Day, Cats were near at hand for the rescue. Many times Cat pilots attempted rescue even if the downed aircraft remained under enemy fire. Several Cats and crews were also lost during these operations.

Japs showed no mercy and made a practice of shooting airmen while drifting down in a parachute. They would also race rescue aircraft to the site of downed wreckages in an attempt to kill or capture surviving air crews before they could be recovered.

B 17 WITH DUTCHMAN

The B17 or Flying Fortress was used throughout the war on bombing missions in both Europe and South Pacific theaters of military operations. It had four Cyclone 1200 hp engines, a top speed of 287 miles per hour, a ceiling of 38,000 feet and a range of 3,400 miles. As you can see, the range of a B17 was almost 1000 miles more than a PBY. Each of the Cyclone engines developed 1,200 horsepower, each burning 50 gallons of fuel per hour at cruising speed.

This adds up to 200 gallons of fuel the B17 used per hour in the air. To fly 15 hours the aircraft needs tanks containing at least 3,000 gallons of fuel.

Grey T. Larison

While not an inexpensive flight, the B 17 made possible rescues over extremely long distances. This made it the best choice when searching far at sea where an extended range capability was essential. All guns, ammunition, and the Norton bomb sight and the bomb bay doors had been removed from the B17's used in our rescue work.

A boat, called The Dutchman, was fitted tight against the fuselage held firmly in place with wire cables attached to the bomb release in the bomb racks, remaining inside the B17. The concept of using a Dutchman on a B17 was not started until the end of hostilities. All available bombers were used to hit Jap installations. When the B17 was no longer in high demand for bombing missions, some of them were fitted with the Dutchman and began use as a rescue aircraft.

The 2nd Air Sea Rescue Squadron had been headquartered on the island of Morotai during the hostilities and moved to Clark Field September 1945, after Japs had signed surrender documents. The 2nd Air Sea Rescue was used for the rescue of any downed aircraft, military or ships in distress.

When the move was made from Morotai to The Philippines it was decided to replace all battle damaged aircraft with new ones recently flown from the States. At the time I arrived on Clark Field, all Cats and B17s were nearly new.

Four B17's, also newly arrived from The States, were assigned to The 2nd Air Sea Rescue Squadron. These were fitted with The Dutchman that was then loaded with survival supplies and made ready for a drop to ocean survivors.

Our pilots were a very interesting bunch. One might think pilots for Air Sea Rescue operations would have received some special training to accomplish rescue assignments. Not so.

It was the end of the war, the government was discharging combat veterans as fast as they could. Hundreds of pilots had finished their service, waiting to go home. The Military decided the fairest way to choose who could go home first was to assign points to each pilot for the number of combat missions he had flown. Rescue pilots had really excessive amounts of flying time into combat areas. They were among the first to be returned to The States. By the first of the year January, 1946, replacement aircrews, including pilots, were flying most Snafu missions.

Fighter pilots did not see much action after the conquest of the Philippines; they were pilots with too few combat points to be returned home first. Former fighter pilots Lieutenant's Gerry, Wells and Sewell who had been flying fighters, were available to fly Snafu Snatchers Cats.

Imagine this, a pilot has been flying the P38 Lightning, Corsairs or Mustangs, aircraft that fly at speeds nearly 400 mph and put him on a PBY that cruises around 120 mph and can only attain 180 mph top speed. Frustrated?

These pilots tried their best to push the Cat beyond its capabilities. Lieutenant Gerry tried, unsuccessfully, several times to rip the wings off during a dive or sharp turn.

The Cat simply laughed at Lieutenant Gerry and obediently accomplished whatever he asked. Lieutenant Wells flew his Cat close to the ground, often within ten feet or so. This comes with some dangers but a highly skilled pilot such as Lieutenant Wells could do this. At times he flew so close to pounding surf that water splashed up against the hull. He came so close to palm trees he was observed returning to Clark Field with palm fronds caught between wing pontoon and wing. Designers have little knowledge of the gyrations the Cat was put through by these pilots.

Pilots assigned to B17s did not have enough points to be returned to The States first as they too lacked combat points. The long range B29 was used during the last months of the war to make the long flights to bomb Japan. B17 pilots had fewer assignments when the war moved north to Iwo Jima, Okinawa and Japan.

Grey T. Larison

B17 pilots were re-assigned to Rescue work from this group, so flying Snafu's was not such a big change for them.

The B17 was the same aircraft used to bomb Germany into surrender. We used it to save lives not take them. Our best use for the B17 was to make those long flights and searches.

Since in the rescue business it was not necessary to carry a heavy bomb load, more gasoline could be loaded. Additional fuel tanks were installed making longer flights routine. Flights of about 1500 miles from Clark Field to a search area were easily attainable, another 2000 miles could be added to fly a search pattern.

Behind the Pilot and Copilots seat there was a hatch in the floor. When opened, a crawl space led to the forward blister. During the war a pair of 50's pointed forward and a single 50 was pointed from each side to protect the area from attacking fighters to both port and starboard. After all guns and the Norton bomb sight had been removed from Rescue aircraft, the forward blister was empty with plenty of room for moving about. It seemed, a person riding in the forward blister could see all over the world.

Whenever I was flew as an observer (searcher) on a B17, the forward blister was my favorite observation post. With a pair of powerful field glasses one was in a choice place to make the first sighting of downed airmen.

During the war, the hatch cover, or door in the floor of the main cabin could be opened, enabling a man to drop into the ball turret to operate a pair of .50 caliber machine guns. Ball turrets underneath our B17s had been removed leaving only a hatch cover fitted into the circular opening.

Bomb bay doors had also been removed from our B17's. This was done to open the bomb rack making it possible for cables attached to the Dutchman could also be attached to the bomb racks. A Dutchman, in position, could be released very much like a bomb when flying a course over survivors.

When one passed from the main cabin forward to the cockpit area it was necessary to pass across the bomb bay. A narrow catwalk was fitted between four solid metal frames where bombs could be attached. This frame was mounted in a V. When bombs were released the lowest one on the V frame dropped first, followed by the second and so on until "all eggs had been laid". The catwalk led across the open bomb bay. When one traversed the catwalk the view, providing there was no Dutchman in place, was straight down to the ground with nothing but the narrow metal width of the catwalk obstructing the view. It was a nervous time.

The "Dutchman" was a boat 32 feet long made to conform to the contours of the aircrafts fuselage. The Dutchman was a self-righting self-bailing craft. Cables extending from the Dutchman were attached to the metal frame inside the bomb bay. These cables were attached to the mechanical releases where bombs were once fastened. Release mechanisms gave the pilot the ability to drop the boat to survivors in the water below. Dutchmen were used when a search was out of range or if the water surface was too rough for a Cat to land and take off safely.

When the decision was made to drop a Dutchman, pilots would make approaches, flying about 800 feet. At exactly the right moment the Dutchman was released. Intentions were it would land close to survivors in the water below.

Released by a static line, three large, colored, heavy duty cargo parachutes opened to keep the Dutchman in an upright position, enabling it to set on water gently, ready for use. It was covered with a tightly fitted tarp, easily stripped off by people struggling while in the water. Hand holds and a rope ladder made climbing aboard possible, providing survivors were not disabled from injuries.

Dutchman's contained:

Two gasoline marine engines, with fuel enough to travel about 600 miles. Folding mast and sails.

Fishing gear included a collapsible harpoon, nylon string, various lures, with patches of colored cloth and copper wire for trolling.

Food and equipment enough to sustain 10 people for 30 days, consisting of both K , C rations and cans of spam.

.30 cal carbine with a generous supply of ammunition.

.45 cal pistol with both lead ball and shot ammunition.

Medical supplies including morphine, tape, bandages, splints and tourniquets.

Sulfa powder and pills to ward off infections from breaks in the skin.

Gibson Girl (a radio transmitter with 250 feet of steel wire cable fastened to a box kite).

Cans of drinking water.

Several Mae West life jackets.

Survivor vests containing a multitude of useful tools, including a signal mirror, short jungle knife, compass, fire starters and a smaller Swiss knife with multiple tools.

Survival vests were placed in a locker on every aircraft and on the Dutchman. These vests had a zillion pockets in them a Very pistol with several red cartridges. Very pistols shoot flares skyward. Flares resembled an over sized shotgun shell (used to signal rescuers) compass, folding jungle knife, two sticks of metal when rubbed together produced sparks, when caught in flammable material, could start a campfire. A small metallic container containing matches for starting fires. In pockets the vest also contained, water purification tablets, bars of concentrated chocolate, a folding cup, a first aid kit, including a large gauze bandage, a morphine tube, sulfa powder and a small circular needle with surgical thread, essential in stitching a wound.

An interesting item, soldiers were constantly trying to get for themselves, was the survival vest. It was neat to have all those pockets full of "stuff".

Field rations of food consisted of K and C rations. These were issued to soldiers in primitive or field conditions. K rations were sealed in wax covered cartons each packed against most environmental hazards, such as salt water, mud, insects and some rodents. These rations were placed inside the Dutchman and in packs dropped by parachute to survivors.

K rations were also placed on board our aircraft for airmen to eat during a flight.

A day's ration of Ks provided 4,500 calories in the three packs, one each for breakfast, lunch and dinner. Each pack included two cigarettes and several squares of toilet paper.

One might, sarcastically add, all the comforts of home. (Never found a toothpick) though.

'C' rations were conventional canned goods, packed in smaller sizes, enough to serve one man a part of a meal. I did not see any C rations packed in larger cans suitable for preparing a group meal.

A useful rule for finding food in the jungle. Watch the plants monkeys eat. Those plants are OK for human consumption. What if the monkey does not eat? Eat the monkey.

As a Rescue Squadron, we might be called to find any crash or shipwreck site, survey what happened and effect whatever rescue was possible.

The geographical area we were expected to fly effectively was about 1500 miles in any direction from Clark Field on Luzon.

This meant we might be called to fly a Snafu about half the distance to the Mariana Islands, Iwo Jima, China, or New Guinea. This included thousands of square miles of water and hundreds of square miles of jungle.

When a call came to our Operations Officer, the usual procedure followed was to first, fly to the last known site of the crash. To find survivors, Rescue aircraft began flying a pattern or grid.

A grid, might be fifty miles square. Parallel courses are flown the full fifty miles about a mile apart, more or less, depended on actual conditions of visibility, wave height, currents and wind direction occurring at the time.

The first pattern might be north and south across the last known site where the accident occurred.

Similar patterns were also flown over an east west course. A given area was over flown at least twice, sometimes more, depending on conditions. Weather, wind, visibility, water currents and terrain were all factors considered to determine the density and time needed for completion of a search.

A Snafu area might be enlarged when searchers were sure survivors were not to be found in the initial search area. Each Snafu was always allowed enough time to make sure survivors could be found, providing they were alive. Several aircraft, including those from other Squadrons, might be alerted and called into an active role, while performing Snafu's. On one occasion we secured help from the 313th Bomb Wing, the B29. They always flew much too high over the sea, couldn't actually sight survivors from their altitude, but they could see a lot of ocean.

A major difficulty with water Snafus was that survivors were rarely found at a crash site. Water currents, tides wind and wave action could carry a floating raft many miles within only a few hours. As days passed, locating survivors became more improbable. The years during and immediately after WWII seasonal changes in currents, wind speed, tides and their effect on floating objects had not been well studied and were poorly understood. In remote areas of the seas the existence of natural currents were scarcely recognized. The exact direction, speed and seasonal changes were barely known nor had they been plotted. With these conditions making an estimate of a survivor's position at any given time on the water, almost impossible. Finding survivors after floating a few days became mostly a matter of luck.

It might seem, sighting a floating yellow life raft on the open ocean is easy. Not so. Waves were generally high enough to "hide" a life raft in a trough created by waves. Searching aircraft must fly almost directly overhead within a few hundred feet of the water, for someone on board to be in position to see it. Flying at higher altitudes so observers could 'see' more water never resulted in faster sightings of survivors floating on the sea.

When survivors were sighted, their immediate needs were assessed as soon as possible and a decision made regarding what we could do to affect a rescue. If a B17 found survivors in the water far from the nearest land the proper procedure, drop a Dutchman.

If the flying distance was within the range of a Cat and the water not too rough, a Cat might be summoned, making a water landing to pick up survivors. Whether a Cat was called or not depended on the condition of the oceans surface or how rough the sea was.

It was extremely dangerous for a Cat to make a safe landing or take off if the swells were more than four feet high. Only highly skilled, experienced and exceptionally bold pilots will attempt it. Some of our pilots, including Lieutenant Gerry, Lieutenant Wells and Captain Barnes successfully made water landings on waves as much as ten feet. However, most of those landings resulted in structural damage to the Cat, at times not repairable. Unfortunately the open South Pacific was generally too rough for a Cat pilot to consider a water landing. Only exceptionally skilled pilots accomplished landings with waves over five feet trough to crest.

Flying Snafus over land covered with tropical jungle was equally difficult. It was almost impossible to sight a wreck through the forest canopy of tropical jungles, even if the fallen aircraft had crashed, exploded and burned. Forest canopies can simply straighten up or grow to cover an entrance hole within a few days.

Searchers look for a slight glint of metal that might be visible for only a split second while passing overhead. Of course, most times even if this tiny glint of sun reflecting from a piece of wreckage actually existed, it might not be observed by anyone on a search plane, no matter how alert or observant.

Yes, men on Snafu aircraft were very dedicated and knew very well the difficulty of the task and the meaning of failure.

Men of The 2nd Air Sea Rescue often flew to exhaustion, making sure their assigned aircraft joined a Snafu for flight after flight, determined to find crash victims. When it was so difficult, or impossible, to visually contact survivors or sight a wreck, not all crashes whether in a jungle or open water are found. Only after every effort had been made and every possible area searched not once but several times was the Snafu canceled. Failure meant no smiling faces for the next several days. Pilots, crewmen and volunteers alike felt the pain of loosing life.

When a crash site was located the next determination to be made, is anyone alive? In the instance of a jungle rescue needed supplies must be dropped by parachute.

Cargo chutes were color coded (White, Red, Green, Yellow, Blue, Brown) enabling recipients to readily identify the contents from a distance. Attached to the shroud lines, in place of a personal harness, a 100 foot drop line might be attached with cargo attached at the lower end. Equipment attached to the drop line could fall unhindered, through the forest canopy, placing supplies within reach of a survivor on the ground.

This was designed so the floating parachute would catch in the canopy of the jungle forest the long drop line would allow the cargo to fall through vegetation stopping either on or near the ground where survivors could reach their supplies.

After they received the radio dropped to them and contacted aircraft flying overhead, a decision could be made whether survivors would be able to walk to a point where they could be given transportation or conversely would it be necessary for a ground party to hack their way through the dense jungle to the crash site.

After pickup was made, the manner in which survivors were treated in 1946 was nothing like we see today on television. In the mid 40's rescue aircraft could only SNATCH AND RUN taking a victim to the nearest hospital.

Once a survivor was taken on board one of our aircraft we had very little means to stabilize the person. Many times there was not a Medical Technician or Corpsman, one who was well trained in treating field emergencies, present.

Our standard procedure was, strip survivors out of wet, torn, dirty or bloody clothing, wash them using a "whore's bath" then wrap them in dry blankets and give them a warm drink such as coffee or hot chocolate, sometimes followed with some hot food.

We stocked only tubes of morphine, sulfa powder and pills, bandages, splints, disinfectants and blankets.

In the mid section on the CAT, there were four bunk beds, attached to cabin walls, where survivors could be made comfortable. On the B17, there were no bunk beds. However, there was considerable floor space available where portable litters, stored aboard, ready for use when needed.

When rescue aircraft reached Clark Field, patients were loaded into a waiting ambulance, waiting on the airstrip and whisked three miles to the base hospital. Adequate care of the day was available there. However, considerable time might have elapsed between the time a survivor was sighted and his entry as a patient into the hospital. There were times when the flight time from the pickup site to the hospital at Clark Field was more than 6 hours. It's probably needless to point out some survivors were lost because of the lack of adequate "right now" emergency room treatment.

The Squadron Area

In the Squadron area tents, standard pyramid shaped with a four foot wall, mounted on a wooden floor were used for sleeping quarters. Inside four cots had been placed, one in each corner, with sufficient space in the center for a table. A locker for personal items was placed at the foot of each bed. Previous occupants of this tent had fastened a large tarp overhead, stretching from each of the four corners to cover an area the exact size of the wooden floor. The tarp kept some intense tropical heat trapped above the heads of standing people. Blocking heat overhead allowed gentle breezes to keep air moving in the living area. The tarp also caught rain mist seeping through the canvas tent. Collections of water falling into the tarp were channeled outside the tent.

Swale type buildings with a poured cement floor were used for the Mess Hall, PX Operations, Mail and Orderly Room where the Squadron Commander and Adjutant had their offices. The Quartermasters Supply Building, Post Office, Latrine and Motor Pool were housed in other swale buildings at the opposite end of the Squadron area.

There was a separate tent area where Officers were housed. The Military practices a strict caste system. Officers and enlisted men must never socialize or be together when not on duty. So of course, Officers never ate in our Mess Hall. Where they ate I really don't know. It must have been at an Officers Mess located far from the Squadron area.

This always seemed so silly to me, as these men both Officers and enlisted men, placed their very lives in each others hands every day. While flying a Snafu, 'Sir's' were muted. Rank was acknowledged when addressing anyone of superior rank but sometimes, in tense situations, first names were used. This is the way our men regarded one other when there was a Snafu to be accomplished.

Allow me to introduce you, the reader to my tent mates. S. Sergeant. Jimmy De Lisle used the cot immediately to the left when entering the door. He was a well trained parachute rigger, a very responsible and conscientious soldier from San Jose, California. Jim was married only a few days before he left for overseas Military Service. He and his wife Gerry were young and very much in love. Jim talked with me about Gerry much of the time. Honestly, I was also talking to him about my wife, Ginny, left behind in Odessa, New York. We exchanged homesick feelings, plans and fantasies.

Using the next cot was a soldier I never did get to know or his name. He had a 'leave me alone' personality. It's a mystery to me what his work was. I guess he was a mechanic, whether assigned to a B17 or PBY I never really knew, even though he slept near me several months refusing to share his life with anyone. He disappeared during non-working hours, never lying in his cot napping. He did his job as scheduled, returned weekends, sober and was never detained after "short arm" inspections.

The man occupying the cot in the far right corner was Technical Sergeant. Bruce Olson, a Crew Chief, an extremely responsible job, assigned to one of Squadron's B17s. The Crew Chief is charged with keeping his assigned aircraft in perfect flying condition, if anything goes wrong mechanically while the aircraft is in the air, the Crew Chief is held responsible.

Bruce and I became good friends after talking about what we might do after discharge from service. Bruce wanted to build and operate a fishing-hunting camp on one of Wisconsin's northern lakes. He had a definite area in mind, a place where he had spent joyful summers as a youngster. We talked about building a camp together for fishermen and hunters, passing many pleasant hours making plans we both knew would never materialize. There were dreamy conversations, making us feel good, assuring us there would be life after military service.

The mail clerk made a daily trip to a central receiving center on base each morning around 1000 hours to receive Squadron mail. He would bring it to the Mail Room in our Squadron area, sort and have it ready for distribution at noon when maintenance personnel returned to the Squadron area from the flight line for lunch.

Most of the people in our Squadron worked on our aircraft parked along the flight line. Line personnel were trucked between the Squadron area and the Flight Line, to begin work at 0800 hours. The noon break began at 1200 hours, transportation was provided for return to the Squadron area.

It was a two hour break allowing the men time to wash, eat an unhurried meal, visit the mail clerk and maybe take a short nap in the heat and humidity of the day before returning to the flight line at 2 pm or 1400 hours military time. The work day ended at 5 pm or 1700 hours. This was the daily schedule providing there was no Snafu emergency. In that case every member of the Squadron was to be on duty continuously until the Snafu was resolved.

An outdoor shower had been erected in an area near our tents. Water was pumped up into four, 55 gallon drums, mounted on a steel frame, about 10 feet above the ground. Hot tropical sun heated the water close to boiling temperature, giving us unlimited quantities of hot water, available on demand.

Water was trucked daily to the Squadron area, then pumped up into the 55 gallon drums. Heated water flowed from the overhead tanks to shower heads by force of gravity. Water pressure coming from the shower head was not very strong. Warm water felt ever so good as it trickled over a tired, sweaty body!

The mess hall was of new construction, finished only a few weeks before I reported for duty, was a larger building constructed with swale, a tin roof and concrete floor. Inside, was located separate the dining and kitchen areas, also there was a separate dining area for the Filipinos and a locked storage area for canned foods. A walk in cooler referred to as the "reefer" provided for the short term storage of perishables. Refer temperatures were kept near 38 degrees.

This temperature keeps meat and vegetables fresh a few days until it is served to the troops. Each day we drew food supplies from the Quartermasters. Perishable items were placed in the reefer. Canned goods, for instance cases of spam and C and K rations were placed in the secured locker room. Also, in storage were quantities of canned Australian mutton.

The cook stoves, called a field range, were similar in operation and looked like overgrown Coleman camping stoves. The oven was inside a cabinet, about four feet tall with a front door. The burner, operating on gasoline, could be moved to several positions, depending on what the cook was preparing. The lower position heated the oven. The highest position was located immediately under the griddle. Field ranges were simple arrangements, but very effective and easy to use under field conditions.

The mess hall was set up like this. There was one serving line far from the entrance. Food was served to the soldiers by either one of the cooks or Filipino K P's. The quantities of food served were consistently medium sized. This did not satisfy everyone's appetite as nutrition requirements vary from one person to another. While a soldier could return to the serving line as often as he wanted during each meal, each trip meant again standing in line.

As a soldier entered the mess hall for the noon meal, three times a week, one of the GI cooks, handed him an Atabrine tablet and made sure the soldier swallowed it. Atabrine is not so hard to take as the pill is coated and easily swallowed.

Actually most soldiers resistance to the pill was mostly a matter of humiliation over loss of personal choice. Atabrine does taste bitter if allowed to dissolve in the mouth. Swallow it quickly with a big drink of water, the taste will not be noticed. The only effects noted, Atabrine, will turn one's skin to a light shade of yellow if taken on a regular schedule.

Atabrine was more effective than quinine when used to control symptoms of malaria, a disabling disease; foreign to the typical American GI, was endemic in the Philippines. The anopheles mosquito, an insect that carries the disease, comes calling after the sun goes down. They swarm on a sleeping persons sleeping body penetrating skin with their proboscis to suck a victims blood. Protective netting is a "must have."

The Squadron area was maintained by both Jap POW s and hired Filipinos. They kept the trash picked up, cut grass, washed vehicles in the Motor Pool remove trash and generally did the more menial jobs that always need to be done. No Japs were used in the Mess Hall or on the flight line.

Working in the Mess Hall were about 25 Filipinos, both male and female. They worked as K P's serving food to the G.I.s on the line, keeping the place clean and in good repair, setting up the barrels of boiling hot water for washing mess kits and generally keeping the mess hall operational.

FAVORS NEGOTIATED

During my first few weeks in the 2nd Air Sea Rescue Squadron, changes in personnel assignments had not been made in the mess hall so I was in limbo and out of work. Looking around for some job opportunity it was evident there was a freezer suitable for making ice cream sitting on the patio outside a mess hall door.

There sat an unused freezer and ice cream mixer-freezer. It was what might be described as a forerunner of Dairy Queen equipment. Locked in the storage area there were a number of tins containing powdered ice cream. Ice cream was created from a combination of ice cream powder and dried milk. Additives might include strawberry jam, chocolate powder, vanilla or other flavoring. Any of these were added during the mixing process before placing it in the mixing-freezing unit to stiffen.

This total mix of powder with additives was put in a tumbling freezer. One can see the modern version of this process in a soft ice cream outlet. The partially frozen mix was poured into a #2 1/2-size can and placed in the freezer with a temperature near zero or slightly below to harden.

Grey went into business. Cans of ice cream became "trade goods" for those assigned to the mess hall. Actually very little was served to the troops on the chow line. Instead, it was traded for favors by mess hall personnel, throughout the Squadron.

The PX Sergeant Ed Mastrone sent over a pitcher of 'coke' every afternoon. We would just find it sitting on a table, waiting for us. 35 mm cameras were in very short supply very few coming to our PX. Somehow after Sergeant. Mastrones' visit, mess hall personnel knew exactly when the next 35mm camera would be delivered to the PX.

The guards at the gate were not able to see mess hall personnel returning to the Squadron area late at night or drunk. Possibly they could not see a female being smuggled into the Squadron area.

The First Sergeant in the Orderly room could be very lenient should one of us ask a favor.

One of our people was successful in negotiating for a Jeep taking it on a ten day leave. It did not escape notice, one of the younger, cuter, attractive mess hall girls disappeared from her job for the same ten days. They both reappeared for work the day the jeep was returned to the Motor Pool.

I was to report to the Mess Officer for a regular shift assignment. That ended my little racket with the ice cream. Others took up the position so the mess hall boys were always able to get favors in the Squadron.

Mess Hall Operations

The nine of us GI's poorly trained at Scott Field were assigned to cook in this mess hall. None of us had any idea of how to go about preparing a meal for 250 men, yet military records showed we had been given MOS (specialist number) numbers that indicated we were qualified cooks.

While Filipinos had been doing the cooking since the U.S. Army returned to Clark Field, the military had decided to have the Filipinos do only KP work, GI s were to be cooks. Three assigned to each shift. The older Filipinos had worked for the U.S. Army prior to the Jap invasion and had plenty of experience in meal preparation for the army. All lived with their families in a village on Clark Field located not far from our Squadron area. Filipinos living in one of the on base villages, had jobs somewhere on Clark Field in one or another of the mess halls.

The crew of Filipinos assigned to our mess hall numbered about 25. All ages and both sexes were represented. Each could accomplish assigned tasks without supervision. Actually they worked so well as a team it seemed we GI cooks were just in their way. Possibly they were all relatives as they seemed obedient and followed directions from their elders as children might.

I was appointed First Cook on my shift. We were to work 24 hours on, followed by 48 hours off. Shifts started at 0300 and ended after clean up of the evening meal.

None of the nine of us, three to a shift, three shifts, had any idea of how to run a kitchen. Since appointed First Cook of my shift and felt some responsibility to adequately feed the Squadron of men I went to the head Filipino, an older man and said. "Guno", we have no idea of what we are doing, why don't you and your countrymen continue preparing food in the kitchen as you have in the past. We will learn from you, how to feed these soldiers".

Before he departed for home and discharge I asked the Mess Sergeant who had been supervising Filipino's working on all three shifts for his advice. His answer indicated the depth of his kitchen knowledge. He replied, "When they are smokin' they are cookin', when they are black they are done". That was the extent of his help to us replacements. "Thanks a bunch guy you have been a big help. Now go home, we will deal with the Filipinos and Squadron personnel in serving acceptable meals".

A section of the dining area had been set aside for the Filipinos to eat their meal. There were about 25 of them working in our mess hall. Their numbers varied from day to day, why, was impossible to understand or did it make any difference. The Filipino crew, regardless of numbers, accomplished their work on time every time with little supervision from any of the G.I. cooks.

Some were accomplished cooks, others were the KP's doing the menial tasks like serving on the line, cleaning or materials handling around the kitchen. Actually the Filipino cooks, older more experienced people directed the KP's. Filipinos who worked at other locations within the Squadron also ate their meals in the mess hall with our Filipino personnel.

Filipinos brought some of their own food although abundant quantities of white rice were available for them from our storage room. Possibly food served GI's was too rich for their digestive systems. They did seem to eat fish preferring fish heads, fresh fruits and vegetables. Filipinos were small undernourished dark skinned people many of them barely exceeding 100 lbs. This was probably because of the Jap occupation when they were not allowed enough food to maintain good health.

The serving line was a low table located at the end of the dining area. When ready for serving, food was taken a few feet from the kitchen area to the serving line in the cooking containers the food was prepared in. Food was served hot from the cooking utensil directly into a soldier's mess kit.

Soldiers would then go to a table of their choice in the dining area to eat and socialize with one another. When soldiers finished their meal they were required to wash their own mess kits outside a side door to the mess hall. (No automatic dish washer as in the States) Three fifty five gallon drums, with the tops removed, were placed on cinder block mounts, lifting them about a foot from the ground.

Burners from a field range were placed in this space directly under the 55 gallon drum, when filled with water, was heated to boiling temperatures. The first drum containing scalding water soldiers passed, contained dissolved strong soap, (detergents were unknown at the time).

The second and third drums contained clear rinse water. All were kept at boiling temperatures by burners placed underneath between the cinder blocks. It was required each mess kit be placed in the boiling rinse water for at least a minute. This requirement was necessary to kill all possible bacteria present as guarding against bacterial infections must be a constant precaution in tropical areas. After washing and rinsing, mess kits were taken by the soldier to his tent where a special rack had been erected to hold the mess kits in the air to dry in the sun. The Mess Officer watched this procedure every meal.

Mess personnel were to draw rations daily from the quartermasters. We were given a supply of those items to be served that day. Everything was carefully weighed or counted to give us exactly enough for soldiers assigned to our Squadron each day. If our head count differed, so did our rations. With our daily rations, we were required to take cases of canned Australian mutton and a few boxes of K and C rations.

The Military actually made plenty of quality foods available. However it was all canned, packed in large containers or held in cold storage during shipment to the Philippines.

Eggs were put in cold storage fresh as soon as they were laid, later loaded on a refrigerated ship consigned to Manila. There they were unloaded and again placed on cold storage.

Later they were trucked to Clark Field for distribution to the various units there. Obviously it took considerable time for the egg to travel from the hen to a soldier's plate. They could hardly be referred to as FRESH.

When we new cooks first started our assignment in the Mess Hall there were dozens of cases of Australian mutton in the storage area. Why was it there and what to do with it? Australian mutton was a problem, as GI s would not eat the stuff. This mutton was from spent animals, not spring lambs. It had a musty taste and smell. Try as we might we could not get rid of the stuff. No one including our experienced Filipino cooks could prepare it with a taste and consistency GI's would accept.

When we discovered Filipinos loved mutton, a deal was negotiated. Cans of mutton were exchanged for fresh fruit and vegetables. Filipinos brought in bananas, citrus, pineapples, lemons, limes coconuts, mangoes and various kinds of nuts. This was all outside regulations. However, Major Jarnigan "Jungle Jim" our Commanding Officer made no move to stop the practice. These fresh items were placed at the end of the serving line, available to all. Each soldier could choose whatever he preferred. Filipinos did not consume ALL of the mutton. It was reliably reported, canned mutton was offered for sale in the Angeles village market. Possibly some was resold and shipped to Manila.

We were also required to take quantities of K rations. Actually, they were not difficult to pass along, as K rations contained several items very welcome to soldiers like cigarettes, candy, concentrated chocolate, meat and music paper.

Rather than letting K rations pile up in the storeroom they were placed in the dining hall where soldiers could freely take whatever they wanted. K ration items were taken to the tent, mostly to be used when leaving the base or when flying. K rations were most appreciated during long Snafu flights as they could also be used to provide quick nourishing food to survivors when taken on board our aircraft.

Packages of K rations were packed in a wax covered paper box about the size of a brick. These were easily carried and were frequently used as a snack during evenings in the tent, while traveling off base to the city or while on flights. Cans packed inside "the brick" contained compressed dried food items.

A stove for heating cans of K rations could be easily fashioned from the soft aluminum containers used to pack K rations. After the contents of a larger aluminum can had been consumed and empty it could be cut down the sides with a pocket knife. Four cuts were generally made about half way down the side of the can. A handful of dirt was dropped in the bottom; a few drops of gasoline were then added to the dirt. Open cans of food, ready to be heated, were placed on the top of this miniature stove. The ever present cigarette lighter ignited the gasoline, moments later; a can of hot food was ready for consumption.

While our kitchen equipment was primitive, we were able to put out quite respectable meals. There were select and prime grades of beef, pork and chicken. Chicken arrived to our mess hall in frozen parts. We could not keep anything frozen so prepared and served chicken the day following receipt. Without a deep fat fryer, most chicken parts were either breaded and fried on the grill or baked. Pork loins were cut into chops and prepared in a number of ways. When a shipment of pork arrived we would have pork several days perhaps several weeks before a shipment of something like beef arrived or the next shipment received might be poultry. Food shipments did not seem to be mixed. Some monotony needed to be tolerated.

Cooks learned to be inventive when preparing our foods to keep them acceptable to the troops.

'A food that never gained acceptance from the GI''s was SPAM. Men repeatedly said they hated spam. Spam came in tins weighing five pounds each. Our storeroom was full of it and we served some almost every day. Spam was prepared in many different ways. We tried various spices and combinations with other foods, especially vegetables, but no matter what we tried, it was still Spam. I don't think it was the taste that turned soldiers off it was more the monotony of having it every day.

Bread of several kinds like white, wheat, or rye arrived every day. While the bread was freshly baked in Manila and trucked to us every day one needed to be watchful for the little beasties who almost immediately after baking could be seen crawling around in a slice of bread. Maybe ovens heat enabled them to hatch them out. Where they came from was a mystery to us.

They were easy to find, just hold a slice of bread up to a light source, beasties could be seen crawling around inside. Just pick it out first if it bothers you before eating the bread, it wouldn't hurt you one way or another, don't worry about an extra or unplanned protein. Remember this protein was free, that is the Military did not charge extra for it.

There were always plenty of fresh eggs in the reefer. While they were fresh, its true eggs were generally transported by ship from The States to Manila. Well, not entirely. We did have aircraft flying to Australia occasionally. On these missions someone was always designated to buy supplies of fresh food, including eggs for the kitchen. Powdered eggs or milk were never acceptable to our soldiers.

We never received dried, powdered potatoes. Fresh potatoes were received as fresh as possible in one hundred pound bags. Since skins were very thin, potatoes were prepared skin on, we received no objections. Potato skins contain nutrition not found in the remainder of the potato.

Sunday mornings we made every attempt to prepare eggs to suit individual soldiers. We had a long griddle about 6 feet by 2 feet. Three burners were placed under it.

When the men came in for a leisurely breakfast we asked each soldier how he wanted his eggs that is sunny side up over easy or scrambled. Someone had taught me how to make an omelet using plenty of graded cheese, ham and sausage. With some practice one can became quite accomplished making omelets to order.

Most GI s, including myself, did not know the secret, how to easily open a coconut. One could hammer on a nut all day and still not be successful. Marina, one of our mess hall workers, took me aside one day to demonstrate. Using a table knife, she taped the coconut along three seams. Continue tapping a few minutes, rotating the nut so all seams have been lightly tapped several times. In a few moments the coconut will crack open, the milk can be saved and the meat exposed. The processes are so simple most anyone can easily open a coconut. Brute strength is useless and not necessary.

Our Squadron Commander, Major Jarnigan ("Jungle Jim") was a combat veteran and not inclined to adhere to the spit and polish of the Regular Army. All soldiers under his command had Class A passes in their possession. Snafu Snatchers could leave Clark Field without first getting a pass issued from the Orderly Room. That meant his personnel could leave the base most anytime they wanted.

The only requirement, we report to the work schedule, on time, every time. There were no Saturday morning formations, that is, marching in circles, doing close order drills on the parade grounds. Nor were there any of those demeaning white glove inspections many Squadron Commanders require. He did insist on superior job performance and for maximum effort from all personnel, when the

Squadron was flying a Snafu. As an occasion might demand, men customarily worked beyond what might be expected of them. A Rescue Squadron with a Commanding Officer like Major Jarnigan inspires men to great achievements.

To the west of Clark Field there are some low-lying hills. A dirt road led past the Hospital and into the hills where one could overlook the airbase. Japs, when defending Clark Field against the US Army, had dug several caves deep into hillsides where they had a clear view and were in position to direct artillery and machine gun fire on American troops as they drove south from the Lingayan Gulf towards Manila. Defending Clark Field was vitally important to the Japs; they must deny Americans the use of the runway.

American infantry troops had assaulted and burned out the caves with flame throwers. Flame throwers are a brutal weapon of war it will kill not only by burning a man to death when hitting him with burning jelly gasoline, it also kills by burning all oxygen in a confined place, like a cave, suffocating anyone trapped inside.

Jap soldiers refused to surrender no matter the odds fighting to the death were demanded by their culture. Japs believed dying in battle was a sure way to gain honor for the Emperor and get to heaven. When Jap resistance decreased sufficiently, the cave entrance including several yards inside the cave was blasted by American troops with high explosives, causing the roof to cave in. It is certain Jap soldiers were buried inside some of those caves and remain there. War is a brutal fight to the death. Doubtless the families of these Jap soldiers never knew how or where their loved one died.

The hills extending several miles to the west from Clark Field were covered with dense tropical rain forest. It's impossible for a person to penetrate jungle growth of this kind without extensive cutting with a machete. Hidden in this forest jungle lived a group of Filipinos known as Negritos. They were small people, with a much darker skin than other Filipinos. They greatly resembled African pygmies, naked except for loincloths carrying primitive bows and arrows. Groups of four or five men appeared on Clark Field from time to time, they never included any women when visiting soldiers.

According to legend, Japs could not continuously occupy the hills above Clark Field. When they sent patrols into the jungle hills, they simply disappeared. No trace could be found. GI's liked to spin yarns about the Negritos being cannibals, this of course was not true, but it is a scary story to tell to newly assigned naive soldier boys fresh from The States.

After loosing many soldiers without gain, the Jap Commander stopped ordering additional soldiers into Negritos territory. Japs were successful in occupying only the highlands directly overlooking Clark Field. It appeared cruelty against Negritos with its accompanying firepower; the mighty Jap Army could not bend these primitive people to its will.

Bows used by the Negritos appeared to be made of a tree limb. The bows were not balanced, one limb being shorter and thicker than the other. The maker had carefully scrapped the wood so it was smooth and polished.

It seemed to me no matter how much the owner practiced, those bows could never be made to dependably cast an arrow accurately enough to actually hit a target.

Arrows were straight swamp grown reeds, feathered with mean arrow points. The points looked like death and were very business like. Aluminum had been salvaged from the skin of downed Jap aircraft. The aluminum skin on Jap fighters was thinner and more workable than the aluminum skin from an American aircraft. Strips were cut about six to eight inches long and a half to three quarters of an inch wide. The edges had been serrated and were twisted in the middle. Arrowheads made in this pattern cut one direction when entering flesh. When the point traveled about half its length into the flesh, the twist started the point cutting in another direction, resulting in terrible wounds.

Torn wounds bleed profusely and are almost impossible to control without special medical supplies. It's no wonder Jap soldiers did not return from forays into Negritos territory.

Negritos looking for a way to make some money were willing to shoot their arrows at any target we might designate. It was great fun to watch them hit a large, oversize Philippine penny. We placed them standing upright in the ground about 50 feet from where we were standing. The Negritos successfully knocked all pennies over with their flying arrows.

Every penny hit became the property of the Negrito who drew the bow. It seemed every arrow claimed a penny for the owner. It didn't take long for Negritos to possess all the pennies in our collective possession.

Jimmy De Lisle and I each had a day off. Jim asked me to accompany him to the parachute loft. He had no work assignment and wanted to show me what is involved when packing a parachute. Naturally I was eager to see how it was done. Soldiers who pack and repack parachutes is called a "rigger"

Chutes used by personnel were made of nylon cloth. It was a strong fabric, rip and mildew resistant. The nylon fabric used for personnel was smooth and slippery.

When it was tightly packed against itself inside a chute pack the smooth, slippery nylon made it almost certain the canopy would not snag when the rip cord was pulled. Shroud lines were also of nylon. These materials were the strongest available at the time.

It was required; chutes used in the tropics must be repacked every month as in the heat and humidity of tropical conditions mildew occurs very rapidly. Mildew, inside a parachute pack, can destroy a fabric within only a few days.

When first arriving to the parachute loft, the chute is popped from the pack, and then hung in a tower high enough for the hanging chute and shroud lines to be free and completely stretched out. Fans blew tropical hot air in and around the hanging chute to dry it. Chutes remained inside the tower drying for several days.

For the purpose of repacking, the chute was placed on a long table. Where it was carefully inspected, every panel eyeballed, and seams double checked for any sign or weakness that could lead to failure. Shroud lines were checked, corrections made for the slightest imperfection. Chutes were inspected several times by various parachute riggers to be sure they were in absolutely perfect condition before it was again folded and placed inside the pack. First the shroud lines were separated, folded into foot long lengths, snapped onto place with rubber bands fastened inside the pack. Rubber bands were used because they held shroud lines in place and offer no resistance when the chute opens.

The next step was to carefully fold the canopy placing it inside the canvas pack. As the canopy was folded, generous quantities of powdered naphthalene were sprinkled between the panels. This precaution is taken to retard any mildew growth.

The canvas pack was closed; a pilot chute placed outside the canopy pack. An additional small flap was closed over the pilot chute containing a powerful spring designed to force the pilot chute to open when released. The flap over the pilot chute was held in place with latch pins passing through a closing plate.

The latch pins were fastened to the end of the rip cord. When the rip cord was pulled, latch pins were pulled from the closing plate, releasing the spring inside the pilot chute forcing it to open. The open pilot chute then filled with air dragging the canopy out of the pack.

This was very exacting, meticulous detailed work. If the chute had not been packed properly it probably would not open.

The last thing, Jim did was to sign and date a small book placing it in a pocket deep within the pack. After a chute was used it will be returned to the loft, if possible. Here it was inspected and either repairs made or the chute was to be destroyed if it could not be repaired to specification. The last person to pack the chute is noted and recorded. Several of Jim's chutes were successfully used to save a life were returned to the loft. All had worked precisely as they should, saving an airman's life.

I might add here, the same care used to maintain and pack chutes for people to use, was also exercised when packing cargo chutes. Because supplies dropped to people on the ground might weigh several hundred pounds, much heavier than a man, cargo chutes were made of a heavier, stronger nylon with a waffle weave finish.

SNAFU'S

While never granted flight status, I flew many Snafus over The Philippine Islands. When a Snafu search was ordered every airplane needed as many observers on board as were available. The goal was to have somebody watching the ground from every blister and window. The more pairs of eyes searching, the better chance of finding downed survivors. It was necessary to actually eyeball survivors and crash sites, electronic equipment did not exist. Snafus were difficult as crashes in The Philippines occurred either in dense jungle areas or an endless sea. These were equally difficult environments to find crash survivors.

Any men, assigned to the Squadron, who was not on other duty and were available, could present himself to Operations where one could sign to join a Snafu. Volunteers were issued a harness, chest parachute and a pair of powerful field glasses. There was never any "flight pay" involved; it was strictly a volunteer duty. I availed myself to many of these Snafus. When a difficult Snafu was being conducted, "Jungle Jim" could depend on several to volunteer for this opportunity to be of assistance.

One might ask "Exactly what are volunteers asked to do?" It could be anything. Some searches required all aboard to simply search the ground. Volunteers might be used to return gunfire coming from Communist Huk or Jap holdouts. During one Snafu, it was a dive from the blister of a Cat followed by an extremely strenuous swim into a very angry ocean taking a life saving rope to a swimming airman floundering in the "drink".

How Snafus are Organized

Every day one flight crew and aircraft were placed on "Standby Alert". This meant that crew must be continuously ready to fly at a moments notice. Their assigned aircraft, pre flighted and ready to go. It only needed the pilot to buckle himself in, run the check list and start the engines. Days could pass without a Snafu being called or a distress signal might be received unexpectedly, making an immediate response imperative.

It becomes most interesting when two Snafu's are called the same day. When an emergency situation occurs, the P.A. Announces the emergency, the Standby Crew, summoned to Operations for instructions. Vehicles were ready to transport crewmen and volunteers to the flight line.

It seemed our aircraft were constantly flying Snafu's. That is to say if some aircraft or ship crashed somewhere in or near the Islands. It was our responsibility to find it and rescue survivors. Pick them up if possible or direct other rescue efforts. Searching over dense jungle existing on most of the Islands was slow tedious work. A crashed airplane in the jungle was most difficult, at times, impossible to find.

The first step when organizing a search pattern was to determine as near as possible exactly where the crash occurred. Not always easy, for many times it was a guessing game as to what course the aircraft had been on and where exactly the emergency occurred. In spite of regulations, some pilots were careless about filing flight plans before they took off pilots might not always adhere to their own flight plan.

May Day calls were generally transmitted under stress while the pilot was desperately trying to keep his aircraft in the air as the pilot knew he was losing it and going down. A pilots emotional tension and/or panic communicates to whoever happened to be operating the radio, thus often desperate messages were sometimes garbled. This could mean there was NO last position known.

Generally strategists understand, distressed aircraft when crashing on the water sink within a few minutes. Survivors, if there were any, would be floating on a yellow emergency life raft. Immediately, there exists an added concern regarding the physical condition of people floating on the sea.

There were many factors and conditions to be considered before beginning a water search. It was almost certain any floating life raft would not be found at the crash site. Several natural conditions constantly exist capable of moving a floating raft great distances possibly within only a few hours.

> First: ocean currents in the search area must be considered. Mariners had only a general knowledge of ocean currents around the South Pacific at the time. They might vary greatly from season to season, sometimes reversing direction completely. This was the first guessing game to be resolved.
>
> Second: the speed direction of movement and effects of tides varies greatly between islands.
>
> Third: another factor to be considered was wind direction. Wind speed, direction and consistency must be added to the equation.

Speed in locating downed airmen especially over water was essential in planning rescue efforts. After some time had passed it must be assumed any raft and/or wreckage had drifted and would likely be far from the position of the original crash site.

We might be searching over hundreds of square miles of ocean. Wouldn't it seem easy to sight a yellow life raft, even a bright yellow one, on the water? Guess again. The open ocean is generally rough enough a life raft can "hide" in the trough between the waves and be most difficult to sight.

When flying a search, whether over land or sea, pilots flew a grid. That is to say, Snafu aircraft would fly back and forth over a designated area following a course beside and parallel to the previous path.

After flying the area, say, from north to south, the pilot would fly the same area flying east and west. Sometimes a second aircraft would follow flying a similar course.

When flying a grid over water, search aircraft might be flying as low as 200 feet above waves or at altitudes as high as 2,000 feet. Altitudes flown depended on the conditions of the sea and general visibility at the time.

Providing survivors had a Very pistol with flares, they could shoot one into the sky to help rescuers sight them. Yellow dye markers floating on the water also helped in sighting a raft.

When a plane went down in the jungle, it would generally crash through the forest canopy, fall to the ground and often explode. No one ever saw an airplane caught in the jungle canopy were it might be an easy find. That was never our good luck to make an easy find. When this happened, jungles were so thick and foliage grew so rapidly, the opening made by aircraft falling through trees might close over in a matter of hours.

Searchers might fly directly over a crash site and fail to spot a wreckage. About all an observer on a search plane could expect was a quick glint of metal, visible for only a split second, flashing through the forest canopy as the aircraft passed over.

Luck had much to do with successful sightings. The more pairs of eyes on board, the search plane the better the chance of spotting that tiny, brief glint of metal, or a floating yellow raft.

March 3, 1946
Getting Acquainted with the B17

My first flight on a B17. It was to be an easy Snafu flying north to the city of Loa in northern Luzon. The purpose, to take pictures of abandoned airstrips that dotted that part of the island. Having never previously been on a B17, I was totally ignorant of the tremendous power about to be unleashed from those Cyclone engines. Four 1,200 hp engines wound up to "take off" speed.

Standing directly behind the pilots and co-pilots seat, looking through the windshield gave a great view of the take off. The pilot taxied slowly to the end of the runway, turned to enter the take off runway, set the brakes and waited for clearance from the tower. In a few minutes pilot warned me he was about to throw the throttles.

Having no knowledge of what was about to happen. I just stood there, intending to get the pilots view as the aircraft rolled down the runway, left the ground and gained altitude. When the pilot receives clearance and starts a "take off" he pushes the throttles all the way forward. As the four Cyclones increase RPM's to begin their song, they develop incredible thrust. As the engines increase RPM's to full power, brakes are suddenly released, the aircraft lunges forward almost leaping into the air. The unexpected powerful thrust forward threw me off balance as I did not have a secure grip on either of the pilot's seats.

Grey was thrown backwards landing on his tender butt, hard. Grey learned an instant lesson about taking a seat and using a seat belt for all landings and take offs.

Sometimes volunteers were assigned to a particular blister but generally each volunteer picked the place where we could be most comfortable and could do the best job for a long period of time. I preferred to take a position in the nose blister on a B17. Sometimes while on a long flight we might rotate observation positions.

Getting to the nose blister however required a bit of gymnastics. Immediately behind the pilots and co-pilots seats, there was a hatch in the floor.

The drill was to lift that hatch and drop your body to a lower level. On hands and knees, crawl forward to the nose blister. During the war this blister contained three .50 caliber machine guns, a bombardier and the Norton bomb sight. Guns and bombing equipment had been removed from our aircraft leaving a comfortable area for an observer to serve for hours if necessary.

From these positions, an observer might be successful in making the first sighting of downed airmen. Once an observer sighted something that seemed important, he immediately informed the pilot over the intercom so the proper action could be taken.

March 17, 1946
A Long Taxi Ride Across Open Water

One of my first volunteer assignments was flown in a Cat searching the South China Sea for, we hoped, a life raft with a single survivor. A small training AT6 had come down somewhere in the area we were to search. This Snafu was to be a mild introduction to the rescue process for me. It was a short search as we were able to start within a few hours of the reported crash. Fortunately, that day there were no strong ocean currents or winds capable of moving a floating raft many miles within a few hours. During the first days search we sighted the bright yellow raft bobbing in the waves very close to the position where had come down.

The pilot, when broadcasting his MAY DAY, reported a fire on board. Time is of the essence in a case like this for those on the rescue aircraft have no way of knowing the extent of the survivor's injuries. Were the burns received minor or life threatening? It was certain, the survivor was not responsive, he failed to shoot a flare from his Very pistol or use water dye to help in sighting him as the Cat approached. He lay on the raft nearly motionless. Although the water appeared rough, Lieutenant Kerney, who was new to the rescue business and the piloting of a Cat, thought the ocean did look calm enough to attempt a landing. He felt this emergency must be responded to ASAP, as there was not time to wait for a B17 with Dutchman to arrive on scene. Should it arrive it would be necessary for two or three airmen to stay with the survivor to care for him and take the Dutchman to the nearest port for further assistance.

If adopted, this course of action would result in additional hours or perhaps days to get the survivor to the Medical care he apparently needed.

Lieutenant Kerney landed the CAT first catching the top of a wave near the raft flawlessly. A rubber life raft containing two crewmen was launched from the Cat. Carefully the survivor was transferred from his life raft to the larger rescue raft then, loaded aboard the Cat. Since waters surface was quite rough the port cabin door could not safely be opened, it was too dangerous, as waves would surely flood inside the cabin, swamping the Cat. It was necessary to securely strap the survivor on a litter then lift him up to the blister.

Our survivor was in very poor physical condition. He had severe burns about the face and shoulders, coming from both his burning aircraft and from the relentless tropical sun. His physical condition was critical. In addition to burns, he also suffered multiple lacerations and a compound fracture in his ankle. His clothing was burned; he was dehydrated and very hungry. After loading the survivor aboard the Cat, Lieutenant Kerney's problem was to get the Cat safely off waters surface. He soon learned it was too rough for him to attain the necessary speed for lift off without risking structural damage to the Cat.

When take offs are attempted in rough water, extensive damage to the outer skin and air frame will likely occur. Rivets may pop out of the skin; struts might bend or be damaged in some way, landing gear or hydraulics damaged so a runway landing becomes impossible.

Should the hull be cracked open the aircraft will swamp with water. A water take off under these conditions will be attempted only when no alternative exists.

Lieutenant Kerney elected to taxi several dozen miles to calmer water, which could be found on the lee side of an unnamed island. It was a rough ride, as Cats are not stable in rough water. Lowering the landing gear, while floating on the water, helps to stabilize the Cat, reducing the motion caused by rough water.

It helped but was not sufficient for crew comfort. Most were soon seasick. Finding water, quiet enough for a "take off," took hours of precious time. Our survivor needed to be taken to a hospital ASAP. Lieutenant Kerney found a small island offering enough protection from waves; he could make the "take off". Once in the air, the pilot pushed the Cat to top speed (180 mph) heading directly to Clark Field on Luzon where the military maintained a large regional hospital.

What medical aid could we offer the crash victim during the three hour flight? The survivor was immediately given infusions of plasma and lactate, as he was severely dehydrated, morphine dulled pain for hours.

His lacerations cleansed, disinfected and a splint placed on his ankle. It was an ugly compound fracture, still bleeding. While he was in the twilight zone caused by morphine, we coaxed some food into him and prepared a hot chocolate drink. Since there was no refrigeration aboard any of our aircraft, cold drinks were 'out of the question'. Our patient was covered with a warm blanket because burn victims need to be kept warm, as body mechanisms for regulating temperature are compromised, chills result. This situation dictated SNAFU SNATCHERS could do little but SNATCH AND RUN.

Grey T. Larison

March 20, 1946
Surviving G.I. POW's

It was consistently reported from the wars beginning, Jap soldiers were extremely cruel, inflicting pain and torture for the sheer "fun" of hearing victims scream or watching them die in agony. The opportunity to hear soldiers involved report such torture first hand came while on a Snafu to the very remote island of Bancala, off the southern coast of Palowan. Feb 5, 1946 was less than a year since the Japs signed the formal surrender in Tokyo Bay, September 2, 1945.

A Snafu had been sent to this island after receiving a radio message from one of our military installations near a small village on Palowan. Operations received a message from Filipino troops concerning three G.I.s known to be alive on the island of Bancala. Operations immediately declared a Snafu.

Having nothing to do that day, I volunteered to fly with the air crew. It was to be an eye opener. Lieutenant Gerry, piloting a Cat, left Clark Field at 1000 hours for the approximately four hour flight to Bancala, landing in a small protected bay on the western side of the island.

Filipino soldiers, alerted by radio, knew rescue was coming and were prepared. They placed three G.I.s in a dug out canoe (boat) and were waiting in a hidden cove. Immediately when the Cat settled down in the water and coasted to a stop, Filipino soldiers paddled the GI's in a dugout canoe to the waiting Cat.

With engines cut, Lieutenant Gerry was sitting dead in the water close to shore making it easier to load the passengers aboard. If there happens to be an adequate beach at the landing site, it is possible for the Cat, with lowered landing gear, to roll from the water up onto the shore. However, this shoreline was not suitable for this procedure as it was cluttered with tangled jungle growth.

The door on the port side of the cabin was opened, the canoe pulled alongside. While these three soldiers were able to climb from the dugout into the Cat's cabin, they were in terrible physical condition.

As soon as they were aboard, crewmen stripped them of their tattered clothing. Every inch of their skin was inspected by the Medical Technician for breaks or parasites of any kind. All three were emaciated, practically walking skeletons, and all probably had malaria. The skin on two of the men had advanced cases of jungle rot, a fungus infection that eats live skin. Numerous cuts and bruises were present, some severe. There were no further bacterial infections in any of the skin breaks.

After inspection was completed, the rescued men were wrapped in warm soft blankets and placed in a cabin bunk. There were a small galleys on our Cats; hot food and drinks, especially hot chocolate were made available to all present before take off. It is really unbelievable how welcome a little hot food can be to people who have been living under near starvation conditions.

A supply a supply of food and medicine had been loaded on the Cat to leave as a gesture of thanks with the Filipino troops who had made survival possible for our men. Lieutenant Gerry personally dropped several bags of these supplies into the Filipino dugout that had carried our men to the Cat.

The return flight to Clark Field was one more insult to the survivors. A radio check with Operations on Clark Field advised Lieutenant Gerry a strong tropical storm had moved into the area south of Clark Field. It was passing in an easterly direction across southern Luzon and northern Mindoro, making it necessary to find a route to avoid it. Taking survivors in such pitiful physical condition into the rough air of in a tropical storm was not a good idea.

It was standard practice to completely fill the fuel tanks on Snafu aircraft in case plans must be changed. Experience taught rescuers, situations can change in a few moments. Snafus were equipped and prepared for most any contingency, including a need for extra gasoline.

There were three choices available.

1. Lieutenant Gerry could fly directly through the storm, in such case it would be a rough ride, but doable.

2. He had the option to fly a course ahead of the storm, circling wide around the cloud cover. This would take more flight time, or

3. Lieutenant Gerry could take the Cat to a higher altitude, trying to rise above the clouds into clear air.

Any decision was difficult. He finally decided to fly around the storm clouds. This course, started with a turn to starboard while crossing central Mindoro, passing over the south eastern most peninsula on Luzon towards the Philippine Sea, here Lieutenant Gerry made a shift in course to the north for the next several hundred miles. Flying over the open ocean, he went north until clear of the storm area. He then turned to port, westward, crossing the northern most part of Luzon until he could clearly see the Lingayan Gulf below. He again banked to port, taking a course directly south for Clark Field.

This made the smooth flight the survivors needed. I think all available cigarettes were used and most food aboard was consumed. Crewmen kept the small galley in constant use. However, we were long overdue for our landing. The Cat was flying the last few miles with only fumes coming from the tanks.

Mess Hall personnel had prepared a hot meal and were open, ready to feed this crew and survivors. Two ambulances were ready at the 2nd Air Sea Rescue mess hall to take survivors to the hospital as soon as they had finished eating their meal. They were whisked off to the hospital where doctors and nurses were ready to care for them.

The first question I asked was "How could it happen? How could these guys still be alive, living on their own in this remote island jungle after these many months?

During the initial invasion of Leyte, the Army drove the Japs from Leyte, later Mindanao and Negros islands. MacArthur then moved his military operations to northern Luzon, invading that island from the north. That landing was in the Lingayan Gulf. Jap units on smaller islands were bypassed and left to starve. If they held US POW's, those prisoners also suffered.

It's an old story in wartime situations. US soldiers killed prisoners in Normandy, Guadalcanal and Okinawa; the Germans killed GI's at Malmady, the Russians had a policy to kill most prisoners. Japs killed GI's taken prisoner throughout the Pacific. The reality of war, there are no rules, its kill or be killed.

Asking questions like this, one might get an answer that is not easy to hear.

Five US soldiers taken prisoner by the Japs at the fall of Bataan were shipped to a prison camp on Palowan Island. Contrary to common belief, not all Jap prisoners were included in the famous "death march" after the fall of both Bataan and Corrigidor in 1942.

On December 14, 1944, the Japs called an air alert (we had been hitting them daily for the past several weeks) herding all prisoners into protected trenches. Next, they pumped gasoline into the trenches with the prisoners. Then the gasoline was set on fire, incinerating G.I. and Filipino soldiers. Several soldiers managed to escape the flames by running into the nearby jungle, with Jap machine guns sending sheets of hot lead towards them. Nine managed to escape. Later, four were recaptured and shot by Jap soldiers.

Originally the nine POW's escaped death by fleeing into the nearby jungle. They were forced to live in the hell of a tropical rain forest for many months. It is a miracle any of them survived. Five survivors were eventually found by Filipino soldiers, transported to a safer area and introduced to Mr. Louden, an American who had lived on Palowan several years prior to the invasion by Japan. Mr. Louden came to the Philippines as a Protestant Missionary, remaining with Filipino people in the mountains after the Jap invasion and occupation. He was 73 years of age, yet he personally led the remaining five across rugged mountains, a trek of 50 miles, to the southwestern side of Palowan.

Five G.I.s were taken by friendly Filipinos, fed and given whatever medical supplies available, although likely this was a very meager amount.

When the opportunity was present, the five were taken by boat from Palowan to the island of Bancala where Filipino's hid them from the Japs for several months.

It was thought only a couple of small Jap patrols were on the island at the time. As the story was told by one of these ex-prisoners, they made their way to the most remote area they could find, hid and stayed there out of sight, leaving no clues. Jap patrols prowled constantly.

Eventually they heard rumors received from Filipino's the war had ended and the U.S. Military and the newly formed Philippine Army was searching for hold out Jap soldiers who refused to surrender.

The Philippines are comprised of hundreds of islands. Most are very small, barely large enough to sustain human life. Others are larger, making suitable places for Jap hold outs to remain for months or perhaps years. These smaller islands were by-passed by the U.S. Army, leaving countless Jap soldiers to survive on their own. Most remained hidden in remote places, never revealing their presence to Filipinos also living on the island. On some islands Jap troops had mountains of war material; they continued the war for many months after the surrender had been signed. No, the Philippines were NOT safe.

Many Japs were forced to live off the land, finding food where they could, refusing to surrender, and attacking small groups of soldiers or civilians for the purpose of stealing food, clothing and/or guns with ammunition. They caused death and destruction for months after battles passed to Iwo Jima and Okinawa.

Some holdouts had been left with considerable amounts of ammunition, guns and explosives. If they still had weapons, Japs made a practice of doing as much destruction or killing as many as they could.

Possibly they never were informed or understood the war had ended. Philippine Operations ended, war moved north to Iwo Jima. These Japs were hunted down by Filipino troops and either killed or captured. The few taken prisoner were eventually returned to Japan.

April 2, 1945
A Leader of Men

If a Snafu presented extraordinary hazards, our Commanding Officer "Jungle Jim" Jarnigan answered it himself. I recall one Snafu when it was necessary for a man to make a parachute jump into the jungle. Since the Squadron had no personnel trained to make jumps, "Jungle Jim" went himself.

A Mustang had crashed in the jungle on Leyte. Although the pilot had broadcast a MAY DAY he was off course by a hundred miles. Consequently it took several days searching dense jungles before the crash site was found. It was verified, a P51 Mustang, with one pilot aboard had gone down in this section of Leyte.

The crash site was in a remote jungle on the island of Leyte. It appeared the pilot was alive, since he fired a flare when his wrecked airplane was first sighted. Since he did not use the radio included with supplies it was assumed he was not picking up supplies dropped. Likely he was badly injured or trapped in the wreckage of his crashed aircraft. After some careful map work, it was estimated a ground party needed three to four days before reaching the crash site. Assuming he was badly injured the surviving pilot might be dead if we delayed in reaching him. Time was of the essence.

We flew over and near the crash site but no place to land aircraft of any kind could be found. It was decided a parachute jump would be the only solution to affect a rescue. The 2nd Snafu Snatchers Squadron there was no personnel available who were trained for a jump into the jungle.

Rather than call for volunteers, "Jungle Jim" and his Adjutant, Flight Officer Smith, dropped through the floor hatch in the belly of a B17. As expected, their jungle landing was most difficult. As expected, both men's chutes caught on the jungle canopy, holding them several feet above the forest floor, probably around a hundred feet from the ground. A long nylon rope was fastened to their harness, the end dropped to the ground. Next they released themselves from the chute harness and lowered themselves down the rope, through tree branches to the ground. This was a tricky maneuver but "Jungle Jim" and Flight Officer Smith were skilled in this sort of work.

"Jungle Jim" found the Mustang pilot who remained strapped in the cockpit of his crashed aircraft. He was able to open the canopy enough to fire a Very pistol flare when our search aircraft flew over him. Somehow that single flare was able to successfully come up through the forest canopy so it could be seen from the air. Miracles do happen. That single flare saved the airman's life. It could so easily have hit a tree branch and never rose above tree tops.

Jungle Jim and Flight Officer Smith reached the fallen airman without mishap. Rescuers had made a correct conclusion; the pilot had been trapped in the cockpit of his aircraft. He suffered a compound fractured tibia, an ugly gash along the side of his skill, probably suffered a concussion as the result. He was dehydrated and barely conscious when his rescuers found him.

As "Jungle Jim" told it later getting the pilot out of the aircraft and on the ground was a difficult process, the survivor could do nothing to help himself. The aircraft had not fallen to the ground, but was hung up in trees several feet above the ground.

The pilot was strapped in the cockpit with canopy open. Medical attention was started before he was removed from the aircraft this included plasma, morphine and some bandages. "Jingle Jim" and flight Officer Smith worked together to lift the pilot from his aircraft and lower him to the ground.

It was decided the two of them, should not attempt to carry the survivor through the brutal jungle to a pick up point on the coast. Instead, the Major decided to make a comfortable camp while waiting for a larger ground party to arrive. He called for more camping equipment, medical supplies food and water.

The Major said the greatest concern after finding the airman was loss of blood and shock; these posed the greatest threat to the airman's life. After building a camp and making the survivor as comfortable as possible, the three men settled down to wait for the ground party to reach them. When the ground party, with five additional rescuers arrived the party prepared for a long slow trek to the sea. The surviving Mustang pilot was so badly injured he needed to be carried.

As the rescue party approached the sea and the pick up point, a radio message was sent to Operations on Clark Field, a Cat was dispatched to land at the evacuation beach. Rafts were lowered, paddled ashore and loaded. After all were aboard, it took an additional four hours for the direct flight to Clark Field.

There are so many difficulties in the jungle and there was much equipment as well as the injured man to transport, "Jungle Jim" realized a ground party needed several more people present. He made a mental note, future Snafu ground parties traversing a jungle needed at least ten participants.

This was a valuable learning experience for "Jungle Jim".

Grey T. Larison

April 10, 1946
The Mountains on Mindoro

A memorable incident for me were the circumstances surrounding a C47 cargo plane that had taken off from Nichols Field, near Manila for the pilot to practice one hour of flying time across country. Since he did not return at the end of one hour, it was presumed a problem had developed.

Pilots were required to fly at least four hours monthly to retain their flight status. Pilots not actively assigned to flying duties, were required to make qualifying flights each month. They might simply shoot (practice) landings or they might elect to fly cross country around the islands. Typically "flight plans" are not filed with Operations for short training flights. In this case, it was not known which direction the pilot had taken, just that he was to put in an hours flying time.

The C47 was overdue. Something had gone wrong. There had been no communication with the pilot since "takeoff", making it a mystery where he was or what had happened. The Snafu Snatchers were notified the aircraft was overdue. Since the pilot had indicated to Operations at Nichols Field he was only taking a one hour flight, Operations Officers thought the aircraft should be down somewhere within 30 minutes flying time of Nichols Field.

Since a crash landing on Luzon, where there were many people and the landscape is flat would have been noticed. This area was not given immediate priority for an air search.

It seemed reasonable to assume the C47 might have been flying through the Mindoro Mountains, a distance easily within range of the 30 minute presumed flying time for a C47.

In the northern part of Mindoro there is a range of uninhabited mountains averaging around 5,000 feet elevation with deep, generally cloud filled valleys. This is an area where pilots commonly go to practice visual navigation skills. However, these mountains can be a trap for an inexperienced pilot.

During morning hours the mountains and valleys in northern Mindoro are generally clear of clouds, enticing for a pilot to fly through valleys. However, by the middle of the day, cloud formations begin to fill the valleys. This can be a death trap for inexperienced pilots. If one flies into a deep cloud filled valley it is probable he will hit the side of a mountain before he had time to pull up to clear it. You see, it's not simply the ability to fly the airplane that is needed, in addition pilots must understand weather conditions, the effects of wind on aircraft, and be able to act accordingly.

Was the pilot flying too low and hit a mountain peak? Clouds, thick enough to obscure vision, are often blown by strong winds into these mountains from the China Sea within a few minutes.

After unsuccessfully searching all other possibilities, it was decided all Snafus were to concentrate searching these mountains. This meant a low, slow meticulous search of mountainsides and deep valleys. Each day for hours, Cats flew close to mountainsides, scanning for any sign of a crash site.

The problem presenting itself was the inconsistent presence of a heavy cloud cover. When searching a canyon the atmosphere might be clear with no clouds present. Suddenly pilots might be blinded by rapidly moving clouds. This required the pilot to immediately 'pull the aircraft up' to a higher safer altitude clear of the tallest mountains. Sudden gusts of winds can change the amount of "lift" the airplane receives, making sudden changes in course impossible. Over such rough terrain flying our usual search grid was impossible. The navigator was constantly challenged to determine exactly what areas we had searched and which were not yet examined.

Snafu Snatchers had previously searched several days without sighting a wreck before I was able to sign on as an observer. The day we flew into the mountains in our Cat, the sky was clear overhead. The pilot, that cowboy Lieutenant Gerry, flew low and close.

In a deep canyon, near a sharp bend, we spotted wreckage. The C47 had been following a deep valley, flying well below mountain tops, until he came to the sharp bend.

Apparently, the pilot's vision was severely restricted by clouds or a sudden downdraft of wind caused a loss of 'lift' to the wings making a rapid 'pull up' impossible. The C47 flew straight into the side of the canyon, exploded and burned. The burning aircraft slid down the side of the canyon wall, landing in the stream below. The visible evidence was a long burn scar down the side of the mountain with engines, propellers and a few pieces of metal lying in the stream bed at the bottom of the canyon. It was certain no one could have survived this crash.

Later it was learned four airmen were aboard, pilot, co-pilot, navigator and crew chief. A ground party would be dispatched to recover whatever they could of the remains.

Pictures of crash sites were required, so we set to work. A steep dive passing over a waterfall was necessary, then, fly about a quarter mile straight and level while passing the crash site and on towards a sharp bend or elbow in the canyon. The pilot must then bank sharply to the starboard, while climbing up over the steep mountainside.

After our fifth pass Lieutenant Gerry, the first pilot, decided he wanted to take some pictures himself of the crash site. He gave control of the Cat to the co-pilot, Lieutenant Nelson and made his way through the cabin to the blister.

Lieutenant Nelson made one successful pass a few feet from the wreck and was attempting the second as the Cat started to lift out of the valley. Something did not feel quite right as Lieutenant Nelson banked to starboard while climbing out of the valley. I was standing in the blister on the starboard side where I could see the passing mountainside coming closer and closer. I could almost reach out and touch it. The Cat was not climbing the way it had previously while coming out of the canyon. It seemed to shudder as if it were stalling. Closer and closer tall grass passed until it was just a few feet from the blister. A hit seemed certain and I was on the side of the aircraft that would hit the ground first! There was not time for a quick 'God help me' as my mind took all this in.

Lieutenant Nelson later explained, a strong gust of wind, rising up the canyon, caused him to loose air speed slowing the Cat to near stalling speed.

Mountains in northern Mindoro are covered with tall grass, perhaps four or five tall. As the Cat passed over the side of the canyon wall, we caught quantities of fresh green grass between the pontoon at the end of the wing.

God has touched my life several times. This was significant testimony to His power and protection. I have always acknowledged and given thanks for His presence that day. An event such as this will cause one to be very introspective and wonder. "What is the purpose of my life? Why am I here"?

Obviously, there was considerable tension on the aircraft after that incident. We had sighted the wreckage site; there was nothing anyone could do to help. In addition, we had experienced a nerve shattering situation. Adrenalin soared through every mans body.

There were no survivors; we had completed our assignment, found the crash site and taken official pictures. With nothing further we could do and the episode behind us, we all felt the need for tension relief.

Lieutenant Gerry flew southwest towards The Luzon Sea and flat land. The small village of Poyompon lie a few miles ahead. In the middle of the town, there was a bridge filled with pedestrians crossing a stream. Lieutenant Gerry thought it would be just great if we buzzed the bridge. He circled the Cat, lined up on the bridge and at full speed, 180 mph dove the Cat heading directly for it. When the Cat passed over the length of the bridge we were only about 10 or 15 feet above it.

Standing in the belly blister all one could see as we passed over the bridge was "ass holes and elbows" as people jumped over the side of the bridge into the stream below.

This probably reads as an incredibly stupid thing to do; yes, in most circumstances one might agree. Being a participant, I can certainly understand the intense emotions felt at the time. We had found a crash site where there were no survivors. We experienced a very close personal encounter with death.

Emotional release seemed imperative. Scaring people by buzzing them with an airplane served as a release from this tension. As the Cat passed over the bridge, men in the blister watching events erupted with an explosive laughter. It was the emotional release needed at the time. Remember, it may have been exciting, daring and certainly was annoying to Filipino's, but we did not injure anyone, simply ruffled a few feathers.

"Jungle Jim" invited every person who had been on that flight to an informal hearing with all present to testify about the incident. There was a discussion regarding both Lieutenant's Gerry and Nelsons' decisions. After making the required flights to obtain official pictures of the wreck, Lieutenant Gerry who had been flying as First Officer gave the controls to Lieutenant Nelson also a seasoned and accomplished pilot. Lieutenant Gerry wanted a few pictures for himself.

"Jungle Jim" listened carefully. He did not attempt to intimidate anyone and heard everyone's view of the experience. He did not put blame on either Lieutenant's Gerry or Nelson; they were both well qualified pilots.

His last comment at the meeting "I hope the leaders of that village have no way of complaining to the U.S. Military about the incident. All of you take a couple of days off".

Lieutenant Gerry was one of the pilots who had been flying a Mustang and had not accumulated enough combat points to be sent home for Discharge. He had been reassigned to the Snafu Snatchers to fly a Cat. The Lieutenant did not have the temperament for this transition. Catalina's were too slow and cumbersome for him. To compensate, Lieutenant Gerry was constantly attempting to push the aircraft beyond its physical limits, stressing its flight integrity.

For example: When making a turn with a Cat, the proper procedure most pilots follow is to use both yoke and rudder to make a slow graceful turn, a thing of beauty to watch and scarcely noticeable to all aboard. Lieutenant Gerry, when making a turn would roll the Cat up on its side so it was perpendicular to the ground, and then yank back hard on the yoke. This snapped the aircraft to a 90 degree turn instantly, stressing the wings considerably.

While in this maneuver Cats moaned and groaned. Some of us wondered whether the wings would come off. I have related this experience several times through the years, often people think my story is an exaggeration. One day a former pilot who had flown Cats over the English Channel, responded "Oh yeah, I used to do that myself"

April 21, 1946
A Jungle Trek

A C45 with five aboard went down on an island lying to the east of Mindanao. On the original flight plan filed in the Mariana Islands were listed two pilots, (military) males, and three female passengers. I cannot tell you who the ladies were. They might have been nurses, USO entertainers or perhaps civilian employees.

It was suspected, the C45 crashed on a fairly large island, Bohol, mostly a mosquito infested swamp, with no known human inhabitants living there. Several days passed before an observer on one of our Snafu flights a B17 found the crash site. The unfortunate pilot had apparently become confused and off course, as the crash site was located several hundred miles from the course described in his flight plan.

One might ask, why did so many crashes occur during these years? Were pilots incompetent compared with pilots today or was equipment too primitive? Navigation equipment was indeed primitive by today's standards, much flying was accomplished by "seat of the pants dead reckoning", which in some cases amounted to just guessing. Radios were not always effective or reliable when used to track one's position. Radar while in existence at the time, was sketchy, a questionable tool that could not be depended on to accurately locate downed aircraft, especially if it were far off-course.

Weather reports were always sketchy, as little was understood about pressure ridges, wind strength and direction. In the areas of the Pacific around the Philippines, water currents had not yet been scientifically studied. Heat sensing devices mounted on rescue aircraft had not been developed. This made finding crash survivors "out of visual sight" extremely difficult at times, nearly impossible. Radar while in use was not sensitive enough to find a floating life raft or a person swimming in the water. Finally there were no transponders mounted on the tail of every aircraft capable of automatically broadcasting radio signals should a plane crash.

Whenever the 2nd Air Sea Rescue Squadron came upon a crash site several days old, there was little chance of finding remains dead or alive. Air crews were not blamed when inquiries were made concerning the whys of not finding a crash sooner. "Jungle Jim" assumed everyone had done his best.

It was only an assumption we might find wreckage on this island, Bohol. Since it was a long flight from Clark Field to this area. B17s was used in this Snafu because it can effectively make a long flight to a search area, have enough fuel to fly a meaningful search grid, and then return to base safely.

It was Jim De Lisle who first sighted a red flare rising through the trees where it popped above the forest canopy. We had found them!

Next, Operations must determine how to recover survivors. We dropped supplies by parachute but radio contact with the survivors was not accomplished.

This raised the question about whether the survivors were badly injured and/or immobile for some other reason? Why were they not recovering supplies dropped to them? Worse possible scenarios must be assumed.

After careful consideration by most of the Officers in the Squadron it was decided a land party could cut their way to the crash site in about two days. Because of his recent experience in making a jungle trek, the Major ordered there should be a party of no less than 10 men.

The more people in a rescue party, the easier it was when difficulties arise and unexpected problems were almost certain to occur. We had no idea whether the survivors could travel on their own or what their physical condition might actually be. How much medical care did survivors need? Exactly what equipment must be carried to the crash site? There were a number of unknowns but Grey, possessing an adventurous spirit, volunteered to be a part of this Snafu. While it was expected to be an extremely difficult and very trying experience, suffering was expected. I(n spite of physical discomfort and pain there were never regrets for taking the opportunity to participate in this part of the rescue effort.

Two Cats with all of us aboard landed close to a sandy beach. Landing gear was lowered and the Cats powered up onto the beach. The cabin doors opened, ten men climbed out. Equipment to effect the rescue followed.

Cutting a trail through a tropical rain forest was a nightmare. Jungle growth is about impossible to penetrate as it grows tight and thick. Vines seem determined to not only be tangled across an intended path, they seem to reach out and wrap themselves around your legs slowing further travel. One could expect to stumble every few yards, with falls landing flat on your face.

Vines came in various sizes. Some were quite small and easily cut aside; others were larger to perhaps the size of a man's arm and fibrous or woody, needing several hits with the machete to be cleared away from our intended path.

We lined up in single file, each of us loaded with packs. (While it might be a guess. Probably each person was packing about 60 pounds.) The lead man cut an opening through the undergrowth so we could all follow. It was exhausting work, forcing the lead man with the machete to rotate with the man behind him. When it was time for a lead man change, he went to the back of the line, the next person at the head of the line took his machete and continued cutting. It might take hours of slashing to gain only a few hundred yards. This was brutal unrelenting work, essential to reaching survivors.

High temperatures and high humidity, together with the extreme physical exertion kept water running from one's body in a constant stream of perspiration. This was followed by intense thirst. Soon every muscle in a person's body screamed for relief but in this situation, there was no relief. A man could only continue to hack a tunnel through the dark, thick jungle vegetation, stumbling and falling as he went.

While taking a turn on point hacking away at jungle undergrowth, I came upon an area where plant leaves were extremely large. These leaves were circular about a yard wide. The stem seemed to come from the side; the leaves were shaped similar to rhubarb plants growing in many people's gardens at home.

Most of these leaves were pointed upward so the flat leaf was pointed towards the sky. Possibly the plant grew to this position so it could gather moisture as it fell through the forest canopy. This leaf was so large someone started calling them elephant ears.

The jungle floor was quite dark. With little, if any sunlight filtering through the forest canopy. Actually I found this area to be colorless with little visible wildlife, only plant life seemed to be present, but there were still the little beasties, making life miserable for all of us. It must be described as a dark and dreary place one would rather avoid if possible.

Leeches were just one of the little beasties everyone must watch for. No one knew where they came from, as they seemed to come from anywhere, coming in various sizes.

Most were dark brown, about an inch or inch and a half long when they first attached themselves to the skin. As they gorged themselves with blood they swelled to become two to three inches long. They attached themselves to the skin and buried their mouth parts under the surface. One does not feel them when they bite; they might not be noticed until blood started flowing after the animal dropped off. While more annoying than dangerous. We viewed them as creepy, crawly, blood sucking and repulsive creatures.

All people in the party watched exposed skin of others for signs of them. Periodically, we would remove clothing and ask another to inspect our bodies for a leech that might have been hidden under clothing. A person's crotch was a 'hang out' for them. A lighted match placed against their bottom will cause a leech to withdraw its jaws. Pulling them out, will result in the parasites mouth parts being torn from the head leaving them under the skin, infections will likely follow.

The gloomy jungle contained hundreds or perhaps thousands of orchids clinging to trees. One couldn't miss them. Colorful orchids seemed to brighten an otherwise dreary scene. I saw orchids in a variety of colors ranging from white to deep red, and there were several sizes, in fact the variety of orchids seemed endless to a 'new' viewer not well versed in the subject. There are all sorts of things growing in the jungle that never came to my attention before. It surely sets one's mind to wondering about the multitude of natural things in this world.

In the tropics near the equator there are twelve hours of daylight followed by twelve hours of darkness. Nights in the forest were pitch black, darker than any of us had previously experienced. Totally black on the ground, although the moon might be shining brightly in the sky. The heavy forest canopy blocked out any possible light coming from the moon at night and the sun during daylight hours. One should never attempt to travel in that blackness. Nighttime brings swarms of mosquitoes including the malaria carrying "anopheles". The Philippine Islands were infested with them. Mosquitoes are not the only little beasties making life miserable for people.

Many insects, fungus bacteria and viruses attack people with little or no immunity, causing infections nearly impossible to successfully treat while the patient remains in tropical conditions.

The man in front of me stepped too close to a snake. In an instant the reptile wrapped itself around the man's leg, climbing upwards. Whether it was a cobra, boa, viper or some other reptile was not determined. The victim needed immediate assistance. I happened to be the nearest person. Without hesitation, I swung my walking stick hitting the snake a sharp blow to the head, dropping it to the ground. Another soldier swung his machete, severing the reptiles' head from its body. Anyone alone in such a situation would have been in grave danger. Cobra's or viper's, strike with lightning speed, injecting deadly venom capable of killing a person within a few hours. Snakes typically strike faster than a person can pull away.

We spent one night in the jungle before reaching this crash site. Although we had experienced an exhausting day, no one slept. Mosquito netting was not effective enough for comfortable sleep.

The ground was wet, a level place to lay one's body was almost impossible to find. Sitting or leaning against a tree or one's pack, was the best position any of us could find. We spent hours of darkness, brushing away biting insects while discussing the ugly situation we were in and a bit about life in general. The usual rain did not fall that night, thankfully.

Each step took our party deeper into swampy terrain. The ground was becoming increasingly wet, with puddles of standing water soaking our boots. Soon, we realized, this part of the island was covered with a shallow swamp, there was to be no end to it. Map and compass were again carefully consulted. Yes, we must traverse through part of it. We waded across several streams lacing lazily across flatlands to form this wide expanse of wetlands.

Fresh water streams eventually emptied into the ocean. Here, in abundance, lived the salt water crocodile. I presumed crocodiles find food in shallow wetlands, which might include smaller mammals, other reptiles, or amphibians. They catch unwary water birds, rodents or fish. Just as crocodiles anywhere do, they consider any life form living in or around a wetland as a potential meal.

While crocodiles specimens we sighted were not considered large, most being in the four to six foot size, they could be extremely fast and had the capability to inflect serious wounds on a person. Crocodiles typically lie motionless in shallow water with only the eyes showing above the surface, making it nearly impossible to sight them when approaching. Crocodiles were well hidden if in water and well camouflaged, remaining motionless when approached. They can lie on the ground or in shallow water and not be noticed by an approaching person. When these reptiles are hungry, they will strike anything that moves, large or small, including humans.

Since I had been a skilled hunter prior to my military experience I considered myself skilled at finding wildlife hidden in brush, true, while at home in a Temperate zone, however the tropical jungle is a different environment. Crocodiles in that wetland generally were well camouflaged lying motionless until, prey is within a foot or two before moving. Then it is a lightning like strike, with jaws wide open.

At the very least a crocodiles inflict severe lacerations on whatever part of the body it bites. If untreated, infections surely follow. Crocodiles were sighted but no incidents occurred with any of us in the rescue party.

There were however, plenty of monkeys, several species of them. They flew through the trees from branch to branch. Yes, troops of monkeys were noisy. When we disturbed a troop of them their constant screeching could be quite annoying.

While impossible to accurately count, the best estimate when troops flew through the forest canopy close enough to be silhouetted against the sky might have numbered in the hundreds.

In the gloom of the forest I came upon a cute looking, very small animal which appeared to be a monkey of some kind. He was hanging on a small tree limb close to the ground and was so docile it was possible to pick him up for a close examination. It was very small, fitting into my hand easily. The eyes seemed to "bug" out, his ears looked something like a bats. The color was a light brown. I made a mental note of his description long enough to look up what the little guy's name was.

Later in life, while a student at Cornell University, Dr. Sam Leonard a Professor in the Zoology Dept. ventured to put a name on my find. Apparently it was a "tarrier" not a monkey and not a lemur. However, my little friend was docile and very cute, lovable in a place full of unlovable creatures.

When the rescue party arrived at the crash site we found the five survivors. Since the pilot was far off course, he was not successful in broadcasting a May Day, while later he claimed he had, none were received at any listening station. Snafu Snatchers had not searched this area the first several days as other areas appeared more promising. Ten or twelve days had passed since they came down. Thinking they would be rescued within hours, they had not thought of rationing food or water.

These items had been consumed soon after the crash several days ago. When the rescue party arrived all survivors were hungry, dehydrated and quite weak. We concluded, the pilot was a poor leader, unable to take "command" or make "good" decisions in a stressful situation.

Insect bites covered their bodies. Fortunately, most injuries were minimal. The ladies were more frightened and emotional than the pilot and co pilot, who suffered the most damage to their bodies. The co-pilot had a simple fracture in the right tibia, which the Medical Technician carefully splinted and wrapped. It was fortunate the skin had not broken so there was little chance of immediate infection.

The question needing an immediate answer could the man travel on his own or did rescuers need to carry him through the jungle to the pick up point?

Other survivors suffered bruises, some had deep cuts. Our Medical Technician applied sulfa powder liberally to skin breaks and gave patients sulfa pills. Sulfa was the best antibiotic available in the field, at that time, for combating infection.

Penicillin, while available, was used primarily in hospital settings. However, other than the co-pilot, all could travel with some difficulty without aid. We had plenty of morphine, sulfa powder, bandages, a couple of splints and one litter.

After assessing all survivors' physical condition, it was decided it would be best if the co-pilot were carried on the litter. If the injured man tried to walk on his own it would likely result in further damage to his fractured tibia. Putting his weight on the injured foot in an attempt to walk might turn a simple into a compound fracture. Jungle floors are not smooth paths, making for an easy stroll. Keeping one's footing was a constant challenge as stumbling and falling were impossible to avoid. No, he must not attempt a walk to the sea.

Injuries were immediately given attention, followed by a large hot meal. With only a few hours delay, the return trip began through the jungle to the pickup point on the coast.

A bright experience happened for me while on the return to the pick up point. There, in shallow water, stood a heron in the usual fishing posture. This one looked similar to the great blue heron commonly seen in New York State, my home. This heron was about the same size as a Great Blue.

This heron had black wings and white breast. Yes, there was a sort of beard to complete the picture. While I could never name the heron, either then or later, it gave me a good feeling to see something so familiar.

Early in life my maternal grandmother had interested me in watching birds. She pointed out to me the endless sizes, colors, bills and shapes that could easily be observed. She instructed me, birds had names. Developing in me the lifelong habit of watching bird life.

Perhaps birds were present but they were well hidden in the Philippine rain forest, bird life in this forest seemed very scarce to this visitor who constantly kept looking for birds but 'eyeballed' few.

Treetop birds might have been present but were much too high and hidden in forest canopies to be seen from the ground. Ground feeders, if present, were hiding under dense litter lying on the forest floor. Few shore birds and water waders seemed to be present around the wetlands, possibly because of the presence of the top predator, the crocodile. Plenty of food seemed to be available for shore birds i.e. seeds, insects, grubs, worms and small fish.

The return trip was a repeat of the trip into the crash site, except a path had already been cut through the undergrowth. This passage did not require much additional cutting, making it easier for each individual to travel but, we were burdened with a man on a litter.

Again we rotated the heavy work. This time not swinging a machete but carrying the litter. One of the ladies, an Amazon of a woman, insisted she take a turn carrying the litter, while we tried, she could not be persuaded otherwise.

When we arrived at the shore of the wetlands containing crocodiles, survivor's reaction was almost comical, much different than what one might expect.

The ladies accepted the perceived various dangers without visible distress, completing each task as requested, carrying their full share of the required tasks. The first instruction, stay close to one another, select a stout stick to be a constant companion. In addition to being used as a "walking stick", it might be used as a weapon in the event a crocodile, snake or other unwelcome intruder come too close.

We instructed all survivors, if a crocodile is encountered, hit the crocodile just as hard as possible, on the tip of the nose. Since a crocodile's nose is very sensitive, they will cease aggression and retreat when it is abused.

The men were different. They could conjure up all sorts of ways to avoid the wetlands, all fantasies. Like walk around, build a bridge; wait here for another rescue party etc. Actually the excuse making was comical, for obviously these were 'city boys' with no practical knowledge of how to survive in a hostile jungle environment.

The co-pilot with the broken leg was at our mercy, he could only lay on the litter going where and when he was carried. The pilot was another matter. It took some salesmanship to convince him our men knew the best procedure for traveling through Philippine jungles. Every one of us in the rescue party were enlisted men, there were no Snafu Officers in the rescue party.

With no Officers in the rescue party, this hot shot pilot constantly tried to take command of the situation reminding us HE was a 1st Lieutenant and outranked all present.

Later, in hindsight, we came to realized, this Officer was stressed out and at the end of his emotional limits. At times he was barely coherent and not able to comprehend the situation. As rescuers were untrained in psychology, none of us were capable of accurately assessing his mental condition or helping to ease it in any way. His comments were ignored.

Because we were traveling with five not very healthy survivors, it took an extra day to reach the coast, forcing us to stay another uncomfortable night in the jungle. Supplies of food and water had been depleted; this made the third day in return jungle travel, very stressful for all. Violent physical exertion without nutrition and water soon exhausts even the strongest person.

As we approached the sandy beach, our radio man called Operations on Clark Field. When we finally arrived at the beach, two Cats were waiting to pick us up.

The pilots, Lieutenant Nelson and Captain Barnes, received warm welcomes as all fifteen of us exhausted people piled into a Cat. Every one of us, survivors and ground party alike, were stripped of our clothing and inspected for skin breaks (this is no place for modesty). Then washed, wrapped in dry blankets and offered hot or cold drinks and food. Gallons of water were consumed. Thankfully and with foresight born from experience there was an ample supply of water. Both aircraft had been loaded with additional supplies of food, water and first aid supplies.

Some thoughtful airman, back in the flight line, had placed a case of beer on each aircraft. (He also included several cans of aerosol spray so they could be cooled). All crewmen, working on Snafus had learned to care for others; simple acts of consideration and/

or kindness toward either survivors or each other were common occurrences in The Snafu Snatchers Squadron.

When the rescued survivors arrived at Clark, several ambulances were waiting at the flight line. All, including the rescuers, were immediately taken by ambulance to the hospital.

Survivors were given complete physicals and held a few days for quality food and rest. Rescuers too, were given special attention by hospital staff. Some had nasty cuts and/or bruises all were experiencing extreme fatigue, others were experiencing severe dehydration, exhausted rescuers were held a night or two in the hospital under the watchful eye of hospital doctors.

May 2, 1946
An Engine Change

Technical Sergeant, Joe Cox was married to his job as Crew Chief for a B17. He lived it day and night, working long hours, bullying anyone for needed parts from whatever source he thought might have them. His motivation? To keep HIS B17 in the air responding to all Snafu's.

There were hundreds of war airplanes sitting on the ground that could be cannibalized for parts. For example, to make an engine change, (regulations required engines on every aircraft be replaced when an engine had been run a set number of hours). A NEW or REBUILT engine must be ordered from a supply depot. This procedure required reams of paperwork and a delay of an unknown length of time. Following this procedure could keep a needed airplane grounded for days. The mission of a Rescue Squadron, the purpose of a Snafu, was to find people whose lives were at risk. When a Snafu was called the crew assigned to every airplane must expect to be ready for flight and continue flying everyday until the Snafu was either successful or abandon. Technical Sergeant Cox demonstrated the attitude of the entire Squadron.

It was much easier and quicker to remove a new engine from condemned airplanes, those about to be destroyed. Many of the B17's parked on Clark Field had only been flown from the States, so had a minimum number of hours logged on them.

Utilizing any of these engines was against regulations. It was as if Technical Sergeant Cox and I were of a like mind that is these were pointless and poorly thought out regulations. The government had no intentions of selling these airplanes, they were to be destroyed. Why shouldn't a new engine from one of these aircraft be utilized.

Regulations required a new or rebuilt engine be mounted whenever a required engine change was needed. Rescue Squadrons must have their aircraft ready to fly every day. Engine changes must be accomplished _A_s _S_oon _A_s _P_ossible (ASAP). Grounding rescue aircraft for the convenience of political or any bureaucratic purpose demonstrated the incompetence of some Military and Political leaders.

We were flying a very difficult Snafu; with the crash site far from Luzon, over open ocean between Clark Field and the Mariana Islands. After several long days, the search remained unproductive. Technical Sergeant Joe Cox, a Crew Chief on a B17, arrived back to Clark late in the evening after a long days search. The number three engine on his B17 was scheduled for a change. Regulations require every engine be changed and sent to a maintenance shop where it was to be dismantled and rebuilt when the maximum allowed flight time had been completed.

The usual procedure, to replace engines with new or one that had been rebuilt.

Changing an engine is not considered a difficult process but it does mean setting up scaffolding and getting hoists in place making it possible to remove the engine from the aircraft and lift the replacement to its position on the leading edge of the wing.

Joe stayed with his B17H, his meal that evening was a can of C rations. He called out mechanics, ordered lights be set up. A hoist, replacement engine and scaffolding were delivered to the flight line. He spent the entire night making the change. The procedure of changing an engine on a B17 might take several days. However at 7 am (0700 hours) the next morning, Joe was revving up the changed engine, checking the instruments to be sure all were working to his satisfaction. Joe's B17 with Dutchman left Clark Field with the others assigned to the Snafu.

Had Joe followed strict procedures his B17 would have been out of service several days. Due entirely to his dedication, his aircraft was ready to take part in the search the following day. Joe was aboard serving at his post as Crew Chief. This Snafu included a very long search over endless miles of trackless ocean surface. Every airship available was needed the next morning to resume the Snafu.

I learned later it was Joe, who actually made the first sighting of crash survivors. He had taken a short break from his position as Crew Chief and was standing, with field glasses, searching the ocean out the port window. The raft had been lifted to the top of a wave momentarily when Joe happened to catch sight of it. The Dutchman was immediately dropped to the survivors to make a happy ending.

While Joe slept in a nearby tent and I saw him almost every day; he was not the one to tell me of this accomplishment. I learned of Joe's nights work only when Joe was called to report to The Orderly room. When Joe returned to our tent, he reported to us "Jungle Jim" wanted to discuss matter of the engine change. Expecting a "chewing out" Joe said the Major was completely complementary about the matter and praised Joe for making his decision, insight and courage to take what he considered the proper course of action.

These attitudes are WHY Americans can meet the responsibilities placed on them when they seem impossible. Many former civilians do not respect the straight jacket Military regulations put men in. They simply see a job that needs to be accomplished and do it. Regs be damned.

May 11, 1946
A Call for Help from a Foreign Ship

Some of our situations were very messy and difficult to understand. There were situations, out of the ordinary perhaps, with people we were not familiar with.

Most of our Snafu's were to recover downed military personnel. While most of those distress signals received were from the US Military with survivor's familiar to, us we were also required to answer all distress calls, regardless of origin.

The radio message received was sparse with information, it originated from a ship of unknown origin positioned in the Southwestern Sulu Sea, nearly to the coast of Indonesia. Such a message in a time of war would have been dismissed without taking any action. After the war, distress calls were treated differently. The radio message received at Clark Field stated only the ships position, mentioning one young female had been injured and was in need of medical attention. Our 'Standby Aircrew' responded immediately, taking rescuers aboard a Cat on the Snafu.

The call was traced to a position 200 miles south of the southwestern tip of Mindanao. Our air base near the village of Zamboanga on Mindanao is 700 miles from Clark Field, plus another 200 miles across open water made this a 900 mile, one way flight.

We easily found the ship, using radio signals. As the Cat approached we noted it appeared to be a rusted old cargo type, needing much repair or possibly scrapping. It flew no flag, signifying where the craft was registered. Was this deception?

Fortunately the sea was relatively calm, making it possible for an easy touch down along the lee side of the ship. When the rescue crew, including Grey, boarded the ship, the Captain showed us to a dirty cabin where a young girl lay with a fractured scapula. That was only part of it. She appeared starved, dehydrated and was feverish, in need of immediate help. When our Medical Technician first examined her, the physical condition was so poor she appeared to be dying.

The girl, obviously Oriental, but not a Filipino, probably from another country in Southeast Asia, possibly Malaysian or Indonesian. At first sight, my heart went out to the girl; she was a pretty little thing with long black hair. She was continually whimpering, obviously in considerable pain and certainly frightened.

The Captain of the ship told our rescue team he had picked her up at sea. He admitted having no knowledge about how she was injured or when. The language barrier existing between us possibly explained our lack of understanding but again sometimes the lack of communication is deliberate. Frankly, there were suspicions in this situation. Some of our crewmen had absorbed some "Tagalog", the dialectic spoken on Luzon, but that was of no help when attempting to communicate with any of the ships crew.

Nothing said, gave us assurance we were getting any understanding of what had happened to put the badly injured girl on his ship. We suspected the truth was being denied us. Regardless, we now had a patient very much in need of the medical attention we could get for her. The little girl was carefully wrapped in a blanket, strapped to a litter and gently transferred to the Cat.

Immediately after loading her aboard our aircraft the Medical Technician stripped all clothing from her body, gave her a careful physical examination, followed by a shot of morphine to ease her pain, sulfa powder was sprinkled on her cuts and abrasions. The fracture apparently occurred in the right scapula near the ball joint where the humorous is attached. It appeared to be a simple fracture, with no bones protruding through the skin. Snafu crewmen checked her body carefully, seeking to determine the extent of her injuries. In addition her injuries included, several bruises and lacerations, possibly she had been beaten. In addition she was starved and dehydrated. Morphine soon put her to sleep. After examining her body, the Medical Technician carefully replaced her clothing and wrapped her in a warm blanket. Lieutenant Sewell started the engines, taxied into the wind, threw the throttles, the twin engines began their song and roared into the air.

It also appeared the little girl had not been well cared for while on the ship, as some of the abrasions appeared quite fresh. For sure, she was very underweight, even for a poor Oriental. Oh, how we all wished the Cat could fly faster! Prayers for a strong tailwind were muttered. Nine hundred miles at slightly over 170 mph is close to a six hour flight.

While we were in flight to Clark Field I stayed close to the little girl. She was in need of human touch and compassion. I was one of two available to constantly attend her. The remainder of the crew was occupied flying, navigating and communicating. This little girl was hurting and frightened of what was happening to her.

Long flights on an airplane were not within her knowledge or experience. She had been placed in a bunk bed in the center compartment of the Cat, a relatively comfortable place to be. However, the song of twin 1200 hp engines was not comforting but overpowering and frightening to this passenger, she needed compassion and reassurance.

Our pilot, Lieutenant Sewell, took the most direct course to Clark Field taking us over open water. With the throttles pushed to the firewall, the Cat quickly achieved the top speed of 180 mph. While most of The Snafu Snatchers were young boys, some married, most were single, with no children of their own. One of the crewmen, Rick

Wheeler, had a younger sister about our passenger's age. The sight of this little girl's plight, brought tears to this soldiers eyes. Although he had helped with other severely injured people, this broke his heart when he thought how much the girl was suffering. Rick hovered with me over our passenger the entire flight to Clark.

The information about our patient and our ETA was radioed to Operations at our base. Operations then alerted the hospital on Clark Field.

Had there not been so much tension aboard over the situation, some of us might have noticed it was a beautiful day. The sun was shining brightly reflecting a blue sky from the water. As darkness fell across Luzon, Lieutenant Sewell settled the Cat on the runway at Clark Field. An ambulance was waiting, ready for a quick trip to the hospital.

While in the presence of this little girl I observed complete kindness and loving care given her. Another crewman, the navigator also treated the little girl as if she were his own daughter. There was no negative behavior observed coming from the Medical Technician or any crewman. Even the profanity stopped in the center compartment.

HOSPITAL PERSONNEL WERE ADVISED WELL IN ADVANCE OF THE CONDITION, ORIGIN AND GENDER OF THE PATIENT WE WERE BRINGING AND THE ESTIMATED TIME OF ARRIVAL. (ETA)

UPON ENTERING THE HOSPITAL, A MOST DISTURBING CHANGE IN BEHAVIOR OF HOSPITAL PERSONNEL WAS NOTED.

The remainder of this account is not pretty and speaks poorly of our military and some Americans. How people can hate those of a different color with so much intensity is difficult for this country boy to understand. This patient had never harmed an American and was in need of immediate Medical attention.

This patient arrived at the military hospital on Clark Field in early evening. The Cat carrying her had been pushed to its limits in an effort to hurry her to the hospital. Orderlies took her litter from the ambulance and placed it on a hallway floor inside the door. I decided to stay with her, as the initial reception appeared to be less than satisfactory, I couldn't bear the thought of leaving her alone.

Rick Wheeler remained with me. The two of us stayed with the child all night determined to run interference for her. This little girl was ignored for hours. NO ONE looked at her, this, in spite of Ricks and my repeated attempts to contact someone responsible for newly arriving patients.

In defense of the medical personnel at the hospital I will admit it was late at night and staff was minimal. However, there were people of all ranks walking the hallways past this injured child. During that time period NO other patients arrived, hospital personnel were DEFINITELY NOT BUSY WITH OTHERS, THEY WERE DELIBERATELY IGNORING THIS ASIAN CHILD.

No one gave her triage. Rick stayed with me beside her giving comfort any way possible, getting water and a candy bar for her while stroking her feverish forehead. No one and I mean NO one on the night staff approached this patient at any time. The broken scapula was painful; drugs (morphine) were needed to control her pain. Rick had one last tube of morphine in his possession, we finally decided to take a chance and inject our patient with it, and she was in so much pain. Morphine in injected, the child made no further sound, the whimpering stopped.

Rick and I attempted several times to get the attention of Medical personnel passing through the area, without success. Everyone was "too busy". This, coming not only low ranking orderlies also from high ranking Officers.

After a night full of long and painful hours of neglect, the day staff arrived. With the coming of morning and the day shift of Medical people.

I will never forgive the next events. A nurse, she wore a nurse's uniform complete with Captains bars, a young, very white, red head approached the patient, without comment stripped all clothing from her body, dropping them in a pile at the foot of the litter and walked away, leaving her exposed, absolutely naked.

One could easily observe this child was well into her puberty. Obviously she would be sensitive and embarrassed. This child was a product of what we might consider a "prudish" society, where young ladies stay covered up. How could a trained nurse be so callus and uncaring to not observe and act on this patients pain and embarrassment?

Obviously she did not give a damn about a "gook" child's feelings. Talk about racial and cultural discrimination. How cruel can trained medical people be??? That female Captain earned a DD (Dishonorable Discharge) that night and should be sent home immediately in disgrace.

It appeared this was the mind set throughout the hospital. I wondered "would a colored GI be treated any better in this hospital"???

As a Christian man, it seemed the most humane action I could take, was to do whatever possible to comfort this poor child. I immediately picked up her clothing from where it had been so carelessly tossed on the floor and covered her naked body with her skirt, shielding her from stranger's casual view. Sometime later, orderlies did come after her. I presumed, hoped, they would take her to the emergency room where a Doctor might make an examination, treat her injuries and possibly take X rays of the fractured scapula and place her in a cast. At least that is what I hoped for her.

As the litter was lifted, her little hand reached for me while calling endlessly what sounded like, "Papa, papa", as she disappeared through a doorway. Enough emotion for one night. That cry for help tore my heart out.

Grey T. Larison

Reflecting back on this incident, there is no way I will ever accept the excuse "We were too busy. That patient needed to take her turn." I knew better, from personal observation they were not busy and her turn came and went numerous times.

NURSES, SITTING AROUND THE NURSES STATION WHILE GOSSIPING DOES NOT CONSTITUTE BEING BUSY. This was racial discrimination of the worse most disgusting kind. Sometimes my countrymen exhibit and participate in behavior unacceptable to many Americans.

May 21, 1946
33 Days on a Raft

An Army aircraft, a C46 went down about 300 miles southeast of Zamboanga in the Celebes Sea. The pilot had successfully radioed for help with the May Day signal. His call was answered by a freighter flying the Australian flag. Ships Captain radioed The Snafu Snatchers Operations on Clark Field that he was only a few miles from the crash site and could pick up survivors.

Four survivors of the crash managed to gather themselves on a yellow raft and wait for the Australian ship to find them. The Snafu Snatchers did not immediately put air crews on alert to respond to the emergency, although Operations had plotted the initial crash site in the ocean.

Ocean currents in this area were strong and winds could come from any direction. These powerful forces can push a floating raft many miles within a short time. The cargo ship, lacking any radar, would not be able to find the raft with its survivors if they had been carried any distance from the original crash site.

Reportedly fire caused the crash of their aircraft, a C46. With an uncontrolled fire aboard, Lieutenant Hunter gave the order for all to abandon ship. Four crewmen successfully parachuted out of the burning aircraft. Fires are often followed by explosions as flames reach gas tanks. With this reality, anyone aboard burning aircraft will become very anxious, even panic, in their haste to exit as soon as possible.

With the first hint of danger and the anticipation of an order to abandon ship, all had strapped on their Mae West life vests. The Mae West is a very reliable inflatable life vest that can be instantly inflated with small CO_2 bottles when the survivor enters the water. It will keep a person afloat as long as it remains inflated. Within seconds every man exited the nearest escape hatch.

The C46 exploded while in flight, before hitting the water. One crewman, the radioman, fumbled too long with the parachute and harness. He did not "hit the silk" fast enough. He was still aboard at the time the aircraft exploded.

The four airmen in the raft were the pilot, co-pilot, navigator, and crew chief. When first landing on the water, crewmen were separated by several hundred yards. During the excitement of jumping from the burning C46, someone remembered to toss out the bag containing the deflated rubber raft. It landed near the navigator, Sergeant Durante swam to the bag, pulled the zipper, reached the CO_2 bottle and 'pulled the cork'. The raft instantly popped out of its protective bag, fully inflated. The remaining three airmen swam to the raft and climbed aboard.

Rafts were designed for two people. It could accommodate three if they crowded together. This raft was more than overcrowded with four survivors trying to find space aboard it. There was no food, no water and two very short paddles. They were completely at the mercy of currents, wind, wave and sun with no means of communication with aircraft or a shore station.

Since drinking water is a primary need for people lying in direct tropical sun, Lieutenant Hunter, the Pilot and Officer in charge ordered anything that could be used to catch rain water be collected for use should it rain.

One method of collecting rain water is to spread a tarp or sheet of some kind in a position where it can catch falling water and funnel it to drain into canteens or other containers. It was most fortunate for rain fell every night for the next 10 days, there was plenty of water for the four survivors, at least temporarily. Two of the men had canteens fastened to their belts, they were now put to use. Lieutenant Hunter didn't know a cargo ship was searching for him but knew the May Day distress signal had been successfully broadcast and received.

After searching for two days, the Captain of the cargo ship realized he could not find the floating raft because of strong currents and changing wind velocities. He radioed Operations at The Snafu Snatchers base on Clark Field to advise he was unable to find survivors and needed to abandon his ship's role in further searches. He also reported bits of floating aircraft wreckage had been sighted.

Operations at the Snafu Snatchers base on Clark Field, contacted our Zamboanga Operations where we had previously deployed a single aircraft, a B17H with Dutchman. This was not to be an easy Snafu for the raft was moving rapidly, pushed by wind, wave and water currents also, stormy seas must be anticipated.

The morning of the third day dawned bright and clear. Snafu Snatchers were in the air with first light.

Changeable winds can blow a floating raft in unexpected directions. Wind direction and velocity can change within minutes. Winds might come from one direction for prolonged periods of time. Other times winds can be changeable, making it impossible to determine where a floating raft might be. Since it would not be found at the crash site, broader search areas were ordered. The Celebes Sea is a large empty ocean with few islands.

Sergeant Johannes had broken his left tibia when exiting the burning aircraft, suffering a compound fracture with a considerable amount of torn skin along the broken bone. There existed grave danger of infection and there were no medical supplies available. Sergeant Johannes, in considerable pain, lie on the inflated side of the raft and let his broken leg dangle in the ocean water, claiming it was the only position where he could feel comfortable. Salt water cleansed his open wound, preventing infection.

The bloody leg in the water did cause some concern among his companions about sharks searching for food. It was thought, sharks could be attracted to the smell of minute drops of blood in the water from a considerable distance. Fortunately, throughout the long ordeal, no sharks approached the raft. As long as Sergeant Johannes did not move he remained pain free and quite comfortable. However, continuous exposure to salt water will dehydrate the skin causing further difficulties.

Lieutenant Hunter soon realized the cargo ship answering his May Day signal was not going to find him. Food as well as water was soon essential, methods must be devised to capture any wildlife approaching the raft.

Gulls could be caught by hand. If the men sat motionless a gull might land on a head. A quick movement of a hand and the gull could be caught by the legs. Captured gulls were divided equally and eaten raw. Marrow inside bones was sucked out; anything containing nutrition and digestible was consumed.

The most plentiful catch was "Flying Fish". Flying fish have developed their pectoral and pelvic fins so they can be used as wings and rudder. They search for plankton, their primary food, growing along waters surface. By flapping their pectoral fins, wings, to gain speed, they actually fly short distances across waters surface. They seem to start their flight from the crest of a larger wave and fly to another. Flights have been sighted that measured as much as 50 yards. Flying fish are plentiful in tropical waters, used as food by many Island natives and considered a delicacy. Sergeant Durante had been in the tropics some time and had previously observed techniques used to catch them. Natives used nets; Sergeant Durante used the tarp.

He spread the tarp alongside the raft in such a manner that flying fish might land in it. It was not a highly successful endeavor as Sergeant Durante was a novice at catching flying fish, but his efforts did provide some fish to supplement the gull diet.

However, these efforts did not fill the nutritional requirements of four men. Hunger became a constant companion.

Snafus searches continued without success. Since the search area was so large, Navy and other Army aircraft joined the search. After the raft had floated a week, a Navy patrol plane, not part of the search team, sighted the survivors. An observer on that airplane sighted the raft from a higher altitude. Possibly they had one of the newer radar units aboard making this possible. They dropped a bag, when it hit the water it split open, six gallon bottles of water floated to the surface. There were also three head nets to repel mosquitoes (useless on the sea). A smaller bag contained an essential Very pistol with 6 red flares, also floated on the ocean. These items were successfully recovered.

While not much, other than water, was immediately usable, it raised every person's spirit to know they had been found, rescue would not be long in coming. The radioman aboard the Navy aircraft notified ground stations of the exact position of the raft.

The next day another land based plane flew overhead. They dropped a bag containing goodies including candy bars, more water and several bricks of K rations. They now had some nutrition for several days.

One more test was in store for these survivors. Within only a few hours, a three day tropical storm or monsoon swept into the area grounding all aircraft until it passed. Weather stations estimated the storm would cover the area for three days.

"Jungle Jim" grew increasingly tense about the situation. He knew the position of the raft would change drastically during a tropical storm of this intensity and survivors would be in poor physical condition. He pondered how to speed things up. At Clark Field, available for "take off" were three more Snafu aircraft pre-flighted and immediately available for a Snafu. This included two Cats and another B17H with Dutchman. He ordered them to Zamboanga airfield. With one B17 already on Zamboanga, there were four aircraft available to fly a search once the storm passed.

When the alert was sounded in the Snafu Snatchers Squadron area, I was available and eager to fly on the Snafu. Jim De Lisle talked with his supervisor who agreed to let Jim leave his responsibilities as parachute rigger to volunteer for the Snafu. One other enlisted man, Loyal Wickham, a truck driver for the Motor Pool, also answered the call for Snafu volunteers. This totaled three volunteers for the Snafu, all members of the Squadron determined to contribute to a successful ending. Those men must be rescued!

It was tense and frustrating to receive daily reports of continued failure for one reason or another. While none of us had any special training or rescue skills we hoped we might be able to render some real assistance. Saving lives had become a passion with members of the Snafu Snatchers. "Jungle Jim" gave us his official blessing by relieving all three of us from our regular assignments while we were away on the Snafu.

Jim De Lisle and Loyal Wickham loaded with me aboard a B17H with Dutchman attached, leaving Clark Field early the next morning, Captain Sewell, was the Pilot. Since storm clouds covered most of the southern Philippine Islands Captain Sewell took the B17 "upstairs," flying well above the cloud cover. I wrapped myself in a blanket and slept most of the trip.

Where to look for the raft after this stormy delay? With only four aircraft available for searching, areas must be carefully determined. Again Snafu Snatchers pondered over the maps. Actually, the planning process was little more than a guessing game as the forces of wind and wave were so unpredictable. The search was to be concentrated in waters 300 miles south of Davao on the southern tip of Mindanao.

The morning of the third day dawned bright and clear, again Snafu Snatchers were in the air with first light. Jim De Lisle chose to fly on a Cat, Loyal chose one of the B17s, I the other. Both B17 aircraft reached their assigned search areas about 0900 hours. Flying assigned grids began. While standing at the starboard window with a pair of field glasses I noticed conditions were perfect for the search, the water was relatively calm with unlimited visibility. When we started flying the grid, Captain Sewell flew close to the water, about 300 feet or so. That is about as close to waters surface as a B17 is safe to fly.

Adrenalin was so elevated no one felt tired that day. Everyone was on a 'high' with eyes seemingly glued to the field glasses. Arms usually begin to ache after holding field glasses to the eyes for prolonged periods of time. This day the pain was not noticed. That raft must be found!

Dusk was descending when the B17 flew within hearing range of the survivors. Visually sighting the raft in poor light was becoming increasingly more improbable. At the distant sound of aircraft engines, the B17 was on a course approaching from the south; Lieutenant Hunter fired a red flare with the Very pistol. No observer sighted the first flare. The B17 stayed on a course passing several hundred yards from the raft. A few moments later when the engines sounded closer and the B17, flying another leg of the grid, again came into sight, Lieutenant Hunter fired a second flare.

I was in the starboard fuselage window searching the water below. When the second flare arched the sky, my heart missed a beat or two, my scream into the intercom telling Captain Sewell, the direction of the flare brought the entire crew to full alert. Almost immediately he banked the B17 towards the life raft changing his course 90 degrees to starboard flying directly over the raft, wagged his wings to signal the survivors they had been sighted as he passed over the floating raft.

Captain. Sewell made a pass over the raft to check as best he could the condition of the survivors but it was getting dark so his cursory inspection was not satisfactory. It was noted waters were relatively calm. He then climbed to about 800 feet, lined the B17 on the yellow raft and dropped the Dutchman as he passed over the raft. Three bright orange parachutes on static lines, opened perfectly, dropping the boat, bow first. The Dutchman settled on the sea within a few yards of the survivors.

When the Dutchman was on the water, and the survivors were safely aboard they could indulge in a plentiful supply of food and water.

It was impossible for Sergeant Johannes, the man with the fractured tibia, to climb from the raft into the Dutchman, his companions, too weak to help him. He remained in the raft, clinging tightly to the Dutchman's' ropes. Sergeant Mills, a Medical Technician on the B17 coming directly from Clark Field, arrived prepared for this situation. He had learned via radio there was a badly injured man remaining on the raft after the three other survivors had successfully managed to climb aboard the Dutchman.

Sergeant Mills discussed the matter briefly with Captain Sewell, who agreed to let Sergeant Mills jump. He took the aircraft above 1000 feet where a parachute had time to open before hitting the water. Sergeant Mills snapped his chest pack parachute to his harness went to the hatch in the center of the fuselage of the B17 and jumped, the chute popped open. Sergeant Mills safely "splashed down" on the water close to the Dutchman, daylight was rapidly fading, and darkness was descending.

Without actually examining the three survivors his experience indicated there existed difficult problems, the survivors were too weak to help the injured man aboard the Dutchman.

First, Sergeant Mills pulled the raft alongside the Dutchman and lifted Sergeant Johannes aboard; the man was so weak he appeared close to death.

Sergeant Mills had in his possession a bag containing supplies to splint the broken leg and other medications to treat the victims. These included sulfa powder, more splints, bandages and morphine.

Sergeant Mills opened the supply locker, found a battery operated lantern and started giving his first attention to the man in the most critical condition, Sergeant Johannes. The three other survivors also suffered from a multitude of additional injuries but they could wait a few minutes.

After treating survivors as best he could, Sergeant Mills started the Dutchman's engine. Compass headings were given by radio and a course set for Zamboanga, about 400 miles away. While making this final leg of the journey, Sergeant Mills not only continued to care for the survivors, he also controlled the Dutchman, receiving little help from any of the survivors. They were all too traumatized and weak to offer assistance.

Three survivors were bordering hysteria; so much so Sergeant Mills was forced to administer sedatives in addition to other medications.

Two survivors were so weakened Sergeant Mills found it necessary to hand feed them. All four were in pitiful condition, both mentally and physically. In addition to physical injuries they were severely dehydrated, with considerable weight loss. Exposed skin was covered with blisters resulting from severe sunburns.

Every mile of the way to Zamboanga the Dutchman was escorted by a Cat. A Cat was used for this because it could set down on the water bringing additional supplies, should they be needed.

When they reached Zamboanga after several days in the Dutchman, it was apparent all would survive with immediate care. At Zamboanga Sergeant Mills had access to a wider variety of medical supplies and equipment to treat all four. The survivors were fed a carefully planned light meal, placed on cots in a swale building, given powerful sleeping pills and left to get a good nights sleep.

The next morning the survivors were loaded aboard a B17H for the flight to Clark Field. Before take off, the Dutchman was reattached onto the B17H. When returned to Clark Field, it would be restocked, refurbished and reattached to a B17Hfor use another day.

There remained a very long ride home to Clark Field, the nearest and only available hospital for Military personnel. They would receive excellent care, but the long ride was an additional stress on already stressed men.

It was calculated this raft had drifted 500 miles across The Celebes Sea in 33 days. It's no wonder it was so difficult to find.

It must be noted that Sergeant Mills was not ordered to take the risks he did. He had never previously used a parachute, having received only rudimentary instructions in its use. He saw the situation, knew he could remedy it and acted. He knew his skills were needed and he was available.

Jungle Jim's practice was to let men volunteer for hazardous risks; seldom issuing orders. It was never necessary, because whenever a sticky emergency arose a Snafu Snatcher with the special talents or skills needed to accomplish a mission stood up. Snafu Snatchers frequently achieved the seemingly impossible.

June 5, 1946
Snafu One of Our Own

A call came to Operations from one of our own Cat's. There was trouble aboard. The aircraft had been flying a search along a southwestern shore of Negros Island when a fire of unknown cause started in an area below and forward of the pilots cockpit. Fire had spread into the main cabin where the crew was stationed during flight; crewmen working in this compartment were reported burned. Hydraulic lines controlling the flaps, landing gear and the yoke were disabled. Manila

Bay, the nearest base where repairs might be made, was a flight of several hundred miles. The maintenance shop in Manila was better equipped than the one available on Clark Field to diagnose and repair the damage sustained by this Cat.

Without hydraulics, the controls must be manually powered, brakes, rudder, aileron and flaps were no longer powered, and all must be controlled by manual force. Most important, the landing gear could not be lowered, making a landing on a runway impossible. It was decided the best alternative, attempt a water landing in Manila Bay near an Army dock. Since Lieutenant Barker could not activate the flaps (lowered flaps can be used to assist in reducing air speed when landing) everyone involved understood this landing was going to be a 'hot' one. Wing tip pontoons could not be lowered. Should one of them catch the top of a wave as the Cat settled on the water at a high speed, the aircraft would spin around, causing substantial damage. Air speed when making a 'hot' landing is much higher than is considered safe.

Lieutenant Barker approached the water at around 130 mph. Water landings are usually made with a speed of 70-80 mph when the Cat touches waters surface. If the hull enters the water at high speed a sudden hard jolt will shake the Cat, possibly causing structural damage or the loss of control. Had the water been rough and or a high cross wind blowing, the situation could have turned very ugly.

Lieutenant Barker brought the Cat to the water in a long, low fast approach. Waters surface was almost glassy making a smooth touchdown possible. Lieutenant Barker's landing was smooth, if high speed, landing without incident.

Once on the water and speed reduced, pontoons were lowered to stabilize the Cat. One of the crewmen remarked when speaking of the landing, it was a beautiful touch down, without blemish.

Lieutenant Barker described to me later one of the tricks to making a 'hot' water landing. He explained the best procedure when making such a 'hot' landing is to keep the Cat up on its "step" as long as possible. Friction with the water around the step will slowly reduce the aircrafts speed. When air speed is reduced to less than 80 mph the wings do not receive enough "lift" to remain airborne the Cat will easily settle on water's surface.

As I have stated previously our Commanding Officer, "Jungle Jim" personally took some dangerous or a questionable Snafus himself. Since there was question whether some of our Squadron people had been hurt on that flight, "Jungle Jim" personally took another Cat to meet with our downed aircraft. On Clark Field, he gunned the airship out of its parking area, entered the taxi strip and immediately hit the throttles for a faultless take off.

The aircraft attained the necessary speed to become airborne while roaring down the taxi strip. It is doubtful he had received clearance from the Control Tower for that departure.

Major Jarnigan was generally a careful man, giving great attention to details, but when one of his own was in trouble, the Major focused his mind on what he wanted to accomplish. I wondered since, did he really "get away" with that stunt?' Did the new West Point Commanding Officer have something to say about the matter? None of us ever heard.

"Jungle Jim" landed beside the crippled Cat, transferred the crew to his airship, left a replacement crew to supervise the Cat's repairs at Manila and returned with injured airmen to Clark Field and the hospital. The Major tolerated no delays. Generally a mild mannered man, conversely when situations required, "Jungle Jim" was a bulldog, who tolerated no interference from "Regs".

Since crewmen had received painful injuries and burns his crew were wheeled into hospital doors within a few minutes after landing at Clark Field. Fortunately most injuries were relatively minor and easily treated. After a few days under the care of a competent Doctor and a cadre of pretty nurses, the men were released for return to duty. None of these 'guys' received a ticket to the States.

JUNE 11, 1946
EXTRACTING WOUNDED SOLDIERS

As I have previously mentioned, there was still a shooting war between Filipinos. On the Island of Mindanao, Filipino troops had been in a firefight with Huks. 2nd ERS (Snafu Snatchers) was called to remove three Government troops who had suffered cuts with machetes and/or knives. Since several days had passed from the time the battle wounds were inflicted, until Snafu found them, infections gripped two of the men.

Filipino Army troops were having a difficult time on Mindanao. There was a mountainous area where Huks had a strong point. Here they received regular supplies of arms and ammunition from the Chinese Communists. Mindanao being a relatively sparsely populated island made an ideal staging area for the Huks. They turned the area into a supply and training center. Communism was attractive to poor peasants and an 'easy sell'. Fighting with the Philippine army was inevitable.

The Cat landed on the smooth water of Lake Lanao, taxiing close to shore. As we came near shore the pilot Captain Scott noted a sandy beach near the Filipino soldiers. As he approached the beach he dropped the landing gear so he could roll onto the dry land making it easier to load the injured aboard.

This area of the Lake Lano had several sandy beaches with little woody jungle growth near waters edge. Extending away from the shoreline several hundred yards, the ground was covered with ferns. The largest and most beautiful I have ever seen. They grew up to an average of about four feet tall, covering the shoreline as far along the sandy beach as one could see. Ferns appeared to be limited to this particular environment as none were observed growing in the nearby jungle.

Several species of shore birds congregated along the beach feeding. Present that day were sandpipers, plovers egrets, terns frigate birds, pelicans and gulls. I tried to memorize bird markings so could later look them up in a bird identification book. Gulls had black caps on the head, others had black wing markings.

These were not the ring bills, herring or laughing gulls so commonly present along the east coast of the U.S. Terns hovered over the gentle surf searching for their food lying on the sand as waves retreated. Also present were two species of herons, nothing I could recognize or name. It was impossible to put a name to any of these birds; my knowledge of tropical bird life was so limited. My grandmother would have been proud of me. Maybe she nudged me to take notice of the wide variety of life. I noticed the birds and enjoyed their presence. One could almost forget the danger hiding in the nearby jungle.

The hatch on the port side of the cabin was opened for easy access for the wounded men. Litters were taken outside the Cat to the injured who were loaded on a litter and carefully lifted inside the cabin. Injured soldiers were stripped of their dirty, bloody, clothing, their wounds exposed.

The Medical Technician Sergeant Mills noted deep machete cuts several needed to be stitched together as soon as possible. Sergeant Mills, who had never closed an open wound of this size, recognized the urgency.

He proceeded; first he carefully cleaned cuts on two of our survivors, gathered sutures then to the best of his ability closed those ugly cuts. With "surgery" completed, Sergeant Mills, liberally sprinkled sulfa powder and protected the wounds with bandages. The last procedure, to wrap the patients in blankets and give them hot food and drinks.

When the Sergeant Mills made his examination he concluded some infections had progressed to gangrene. Those victims were in great pain; also gangrene had a repulsive smell. It was strong enough to cause nausea among the crewmen. I might add here, these were experienced rescue personnel, yet nausea overcame some of them. Sergeant Mills did all he could to make the wounded men comfortable for the return flight to Clark Field. He completely depleted his supply of morphine.

Morphine eased pain, blood plasma and lactate boost blood pressure he applied tight bandages to slow bleeding. However, in spite of almost heroic efforts to save him one of the Filipinos with the most advanced infection died on route.

Lake Lano was choked with beds of plant growth growing in shallow water near the shore. As the Cat approached with gear lowered, this plant growth fouled the landing gear. It was determined the gear could not be raised for a water departure because seaweed had wrapped around the wheel mechanism and doors, tight enough they could not function.

Seaweed must be cleared from the landing gear and an opening through the plant growth cleared so the Cat could move into deep water, clear enough, to raise the landing gear. It took some time for crewmen to clear the seaweed from landing gear and open a passageway through the water wide enough for the Cat to pass through without becoming further entangled.

Two of the crewmen stationed themselves on the wing of the Cat with .30 caliber carbines until the water plant clearing process was completed. They were watching for salt water crocodiles. While most of them are not large, the largest seen were about 5-6 feet in length. However, they do pose a threat to anyone in or near water. They can inflict a nasty bite.

Captain Scott eased the Cat back into the water. As soon as the Cat was floating, he raised the gear leaving only the smooth hull in the water. Throttles were pushed to the firewall; twin Pratt and Whitney's began their song. The lift off and flight to Clark Field was uneventful.

June 11, 1946
Extracting Fighting Soldiers

Operations received a distress radio call late in the day, about 1500 hours. This distress message was received from a group of Filipino troops on Palaui Island, located 280 miles northeast of Clark Field and a few miles from the northeastern coast of Luzon. They had been fighting with Huks. Two Filipinos had been killed in a firefight with six soldiers remaining. Four were Filipinos; two were Americans (GI s). Huks had attempted to capture a radar station originally constructed by the U.S. Military but operated by the Filipino Army, two G.I advisors were present. Apparently the radar station had been given to the Philippine government as part of the transition to aid the newly formed Philippine government to gain control of their country.

A Snafu Snatcher was urgently needed. Lieutenant Davis, a pilot with crew for a PBY Cat, was on "stand by" that day. "Stand by," or "Alert" meant the designated crew must be ready for immediate response to a mission or Snafu any time that day. Stand by crews were to be in the air within minutes of the receipt of a distress call.

In areas within a few degrees of the equator there are about 12 hours of daylight and 12 hours of darkness, regardless of the season of the year. Long twilight hours do not exist in the tropics as they do in temperate zones. This would be a very dark night as skies were overcast with storm clouds sweeping across northern Luzon.

This was destined to be a very difficult Snafu. As we approached the prearranged pick up point on the island of Palaui, the hour was late, nearing 1700 hours. That meant it would be dark within only another hour.

The ocean was rough, as usual for the season. Lieutenant Davis felt he could not make a water landing, pick up six men and take off safely with a heavy load. In addition to natural obstacles, there was danger of small arms fire coming from Huks while attempting a rescue from this beach. Lieutenant Davis thought about the factors in the situation, decided the rescue would come off better if attempted from another side of the island where waters surface appeared a little less violent. He then radioed the survivors to hike across a peninsula, a distance of about 2 miles. He would place two rafts, blue side up on the beach where they could be found.

The six on the ground indicated they had received the message and would hike to the pick up point. While the distance was short, along the two miles were other obstacles to be overcome. The jungle was thick, the men had only one machete to cut a path and there was a constant threat Huks would find them. Huks were as vicious in their treatment of captives as the Japs had been. Had these men been captured, they would surely have died violent painful deaths.

Lieutenant Davis flew the Cat to the far side of the peninsula. Crewmen launched the rafts, paddled ashore, placing them where they could be found. After recovering his two crewmen Lieutenant Davis taxied off shore a safe distance.

Again, the sea became rough when winds shifted direction, within minutes, waves rose to the 6 to 8 foot range and the usual high winds were blowing onshore. Numerous coral reefs a few yards from shore hidden by crashing waves threatened the Cat. These could be deadly obstacles should towering powerful waves drive the Cat against them. In addition, a strong tide was pushing the Cat toward the shore. If the Cat hit a submerged reef, one could expect the hull would be ripped open, flooding the interior of the cabin. Extensive damage could result in either sinking the Cat or making a departure impossible.

With these conditions threatening Lieutenant Davis dare not cut the engines, but continued to taxi in circles off shore.

Lieutenant Davis waited for the survivor's appearance and find the raft left for them on the beach. To make the situation more tense, darkness was closing in. The time, now nearing 1800 hours.

The survivors successfully crossed the peninsula and with little delay, succeeded in finding the rafts left for them. All six pushed the rafts through the surf, climbed aboard, and began paddling towards the Cat. Rough water made it difficult to control the rafts; however, the rafts cleared the reefs and approached the whirling propellers of the Cat. The evacuees had so little control of the rafts they feared they would be swept onto one of the whirling propellers. Five decided to jump out of the raft and swim the short distance to the Cat, where they could be pulled aboard.

One man remained in a raft; we learned later, he could not swim. Both man and raft were swept by churning seas below the port wing, missing the whirling propeller, continuing to pass astern of the Cat and out of sight. His rescue could not be accomplished during the brief seconds while mountainous waves swept the raft past the Cat.

This man was doomed to be lost as the raft he was riding in soon drifted far astern of the Cat. After the others were taken aboard, considerable time and effort was expended in a search for him, he could not be located in the darkness.

Churning water was separating the evacuees. Two of the Cats crewmen tied ropes around themselves and dove into the water, at great risk to themselves, attempting to find survivors in the darkness and swim them to the Cat. This was exhausting work. The Cat continued to taxi in circles, searching and picking up evacuees as they found them. One crewman climbed up onto the wing with a powerful flashlight searching waves and troughs. Lieutenant Davis did not dare cut the engines for fear the Cat would quickly drift onto the reefs so apparent during daylight hours. Waves continued to loom higher and higher as onshore winds increased in strength until possibly they were towering more than ten feet, trough to crest.

Although the wing pontoons were lowered into the water, wings were constantly awash as waves swept over them. At times even the windshield was under waves. With a heavy overcast sky it seems there was NO light. A successful "take off" from this very rough sea with high winds was very doubtful.

While no one mentioned it at the time, we all feared Huks remaining on shore continued to look for those six men who had initially escaped them. Had they sighted the Cat with its desperate passengers, it would have been an easy kill for them. Our struggle for survival was less than a hundred yards from shore. Thankfully darkness descended

After loading the five survivors they were able to save, the "take off" was to be another iffy nightmare. This Cat was overloaded with people, making it very difficult for Lieutenant Davis to gain speed enough to become airborne. Waves were smashing the airship, popping rivets from the hull. When large numbers of rivets are forced out of position, the hull will crack open, water starts to flow into the cabin.

Lieutenant Davis had best succeed in getting his Cat off the water, chop chop, right away, or it will soon become impossible. The Crew Chief moved fuel controls to 'auto rich', this gives added power to the engines for short periods of time. He pushed the throttles to the firewall. Before the Cat gained enough airspeed to get up on the step, one final monstrous wave smashed across the bow. It washed across the windshield, momentarily blinding both pilot and co-pilot. This one final wave lifted the Cat to the top of a 10 footer, the Cat "dove" into the next trough. The force of this 'hit' crushed the fuselage severely wrinkled the hull back beyond the pilots compartment to the leading edge of the wing. With so much structural damage, water began pouring into the cabin. Fortunately the Cat became airborne as it hopped into the air from the very next wave.

While it was a relatively short flight to Clark Field, the Cat was barely flyable. It took the strength of both pilot and co-pilot to control flight. They reported the Cat wanted to keep diving; it would not keep its nose up. Keeping it aloft by holding back on the yoke was a man killing job, requiring the combined strength of both pilots.

A safe landing on Clark was nothing short of a miracle as the hydraulics had been damaged beyond use, that meant no brakes or flaps were functional at the time of landing. It was 'hot' and wobbly without brakes Lieutenant Davis rolled the Cat to a safe halt.

Structural damage was extensive as a result of this flight. The Cat, serial number 44-33883 affectionately known by crewmen as "Old 83", was declared a total loss and scrapped.

Lieutenant Davis later commented, he had previous experience with rough water so felt confident he could successfully handle this one. However once on the water he felt differently. This was too rough, he wondered if he really had necessary skills to successfully save these survivors and his crew. Death, waiting on reefs or shore was a constant added concern.

One could easily suspect the devil was trying his hardest to destroy the aircraft and passengers. Was it a contest between good and evil? It always seemed to me WWII was all about this contest. I don't know how else to explain it, because that aircraft with its human cargo and adverse weather conditions should have been lost.

There were so many adverse conditions, but through a miracle, all except the Cat itself were saved. Somehow, Lieutenant Davis found the skill, strength and courage necessary to save these lives.

Emotions run high when one is experiencing miracles almost daily. Yes. While volunteering for these assignments, addiction to the adrenalin "highs" could be felt before, during and after Snafus. Once experienced, waiting for the next Snafu seemed incredibly long. Conviction soon followed, this is why I am here. My personal purpose was not to serve as a cook but to serve on Snafu's during those months it seemed every Snafu had a critical use for me. The mess hall assignment was, the vehicle.

JUNE 21, 1946
A DANGEROUS SNAFU ON ANGRY SEAS

The Snafu Snatchers held an airstrip on Surigao Island near the town of Cagdianao. It was used for the rare situation when a call for help came from the southeastern part of the Philippines. The airstrip had only one Cat with crew assigned to it. Since it was in such an isolated position it had few Snafus to answer. To rescue the crew from boredom, 'standby' crews were rotated almost weekly.

"Jungle Jim" thought it might be good for those assigned to Cagdiano if a kitchen was set up there. I was asked, not ordered, to do the job. "Jungle Jim" gave few orders to his men, a suggestion was adequate. We loaded a B17 with a field kitchen with fuel, an assortment of cooking utensils and plenty of canned foods.

It was estimated it would take about two weeks to set up a mess hall in a swale hut Filipinos would build for us. As small as it was the mess hall was soon ready for business. Plenty of food was available, however it was all of the canned variety. It was not very tasty and certainly monotonous.

Remembering how we traded canned mutton for fresh fruits and vegetables, I ordered a supply of the canned mutton from our storage locker on Clark Field. With these trade goods it was possible to exchange mutton with Filipinos on Surigao Island for fresher foods and to set up an attractive serving line for the air crew assigned to Cagdianao.

A variety of fresh food helped those assigned to this isolated base accept the duty with a more relaxed attitude.

SNAFU SNATCHERS

My tour of duty on Cagdiano had not ended when a radio call arrived from Snafu Snatchers Operations Officer on Clark Field. An Army B24 was in trouble several miles out over the Philippine Sea. The pilot radioed he was expecting it would be necessary to ditch his aircraft. He changed course to fly directly towards Cagdiano, while keeping Operations constantly advised of his position.

The moment came when the B24 could no longer be kept in the air. The pilot ordered his crew to prepare for a forced water landing, hitting the water while 200 miles east of Northern Mindanao. The standby crew waiting on Cagdiano was ordered on a Snafu. Grey, joined this Snafu to make himself useful in any way possible. Within minutes a Cat with full crew of Snafu Snatchers was airborne.

Since there was advanced information regarding the B24s' position, our Cat from the Cagdiago airstrip was over the crashed site almost immediately upon hitting water. During the crash landing, the fuselage was split open, allowing water to fill the cabin. It was already sinking when the crew on the Cat and Captain Barnes sighted it. Survivors were in the water strung out along the last few miles the B25 was in the air. Whatever the problem was, the pilot ordered his crew to leave the aircraft before it crashed and not ride it to the water.

The crew bailed out (jumped) as their aircraft was loosing altitude. In the excitement and tension to leave the stricken aircraft, no one thought to drop a yellow raft. Four men landed in the water, along a more or less straight line extending about a mile across the surface.

Pacific waters were generally bright blue with gentle waves making the water very inviting. On this day dark gray turbulent waters were anything but inviting. As Captain Barnes guided the Cat low and slow, looking for a wave he could catch for a landing. Waves were estimated to be eight to twelve feet high, appearing monstrous, as if seeking to destroy anyone daring attempting to touch them. Crests were covered with angry foam splashing spray seemingly reached upwards trying to capture the Cat and pull it to destruction.

Procedure been decided, 2nd Air sea Rescue pilots would not attempt a water landing if waves were estimated to be greater than four feet trough to crest. Both landing and take off are extremely dangerous if waves are higher. However this situation called for an immediate decision. Waiting for a B17 with a Dutchman to arrive from Clark Field was not an option, as it would take at least 6-7 hours to arrive at the crash scene.

Captain Barnes decided a time delay would allow people in the water to be hopelessly scattered and irretrievably lost. Survivors would not be found as water currents and strong winds we were experiencing will separate and carry anything floating several miles within a few hours.

While he had never attempted a landing on ten foot waves, Captain Barnes felt he had the skills and experience necessary to successfully affect a rescue.

An immediate landing must be made if these men were to be saved. Captain Barnes lined the Cat on the first swimmer to be sighted, came in low and slow against the wind. Captain Barnes decreased engine RPM's, successfully catching the crest of an eight foot wave. The Cat settled into this most dangerous situation.

Angry water poured over the wings and fuselage of the Cat finding its way into the cabin. Every hatch and window was double checked to be sure it was tightly closed, in an effort to keep water out, the Cat afloat and controllable. Once in water, the Cat rolled and pitched, making it almost impossible to keep the aircraft on a straight course towards the position of the first survivor we could see. Captain Barnes dropped the tricycle landing gear, hoping it might help to stabilize the Cat. While it did not turn night into day, dropping the gear was helpful in getting some control over the wild rolling, pitching and yawing.

Finding the men in such rough water became the next challenge with water constantly washing across the windshield it was impossible for either pilot or co-pilot to sight a person swimming and floundering in that angry sea.. Two crewmen climbed to the top of the wing, roping themselves to hand holds to prevent being washed into the sea. From this elevated position they could look down into deep troughs between high waves.

Hand signals were used to signal a crewman stationed in the blister, who could talk directly with Captain Barnes on the intercom, giving him directions, received from the crewmen on top of the wing, to approach survivors in the angry waters.

Fortunately each survivor was wearing a Mae West life jacket and had in his possession one dye marker. A dye marker, when activated, will spread yellow-green dye on waters surface, making sighting easier. However, churning water and blowing wind quickly dissipated dye, rendering the attempt almost useless.

A lookout clinging on top of the starboard wing, sighted the first evacuee. It was incredibly difficult to locate and make a successful approach to someone in such turbulent waters.

The hatch into the port side of cabin could not be opened because waves were slamming against length of the hull. Should this door be opened, water would immediately flood inside the cabin. The remaining option, survivors must be loaded through the blister. Using the ladder from the blister was not practical, as waves would likely tear it away even before use.

Ropes must be used but in this high wind throwing a rope to a man in the water could not be successful. Once a man was sighted, Captain Barnes maneuvered the Cat alongside the man swimming in the water. Crewmen tied two ¼ inch nylon ropes to one of the gun mounts remaining in the blister.

One rope was tied around the chest of one of the crewmen, an accomplished swimmer, who then jumped into the boiling sea carrying the loose end of the second rope. He swam to the evacuee, tied him to the second rope. On signal, crewmen in the blister pulled both men into the Cat.

Next, rescue the second survivor sighted in the water. I considered myself a strong swimmer and part of the rescue crew, I took a turn volunteering for a jump into the sea.

While I did not think such a swim would be easy, success seemed likely. No damn ocean wave was going to stop me! Once in the water, it was almost impossible to swim in any direction, as waves continually swept over my head.

Salt water burned my eyes and found its way into my lungs. For a few seconds I questioned my sanity, volunteering for this violent and dangerous experience. A few minutes seemed forever, as if the struggle would never end. Fortunately it was not a long distance to reach the survivor. Wrapping the second rope around him, I made a quick two half hitch knot in the line tightening it around his chest. We were both safe although we were still in water.

At my signal, crewmen standing in the blister hauled both of us to the Cat's side. We waited in the water for what seemed like an eternity for a big wave to wash us up to a level close to the blister. When this finally happened, it was quite easy for crewmen inside the blister to pull us into the compartment.

Other crewmen took their turn in the water roping the remaining two survivors. I learned to have a little more respect for the ocean. It was a heart throbbing, and stressful experience and a great adventure. While frightened, yes, I would volunteer another time if the situation demanded such a procedure to affect a rescue. It is a spiritual milestone to save another person's life

After all four survivors were safely aboard; Captain Barnes must get the Cat back into the air. The first consideration was to lighten the Cat as much as possible. Everything expendable was tossed overboard. Gasoline was carefully measured and calculated. More than was needed for return to base was jettisoned into the ocean.

Captain Barnes thought maybe if we were patient for a little while these tremendous swells might subside. He continued to taxi the Cat. It didn't take long before all aboard were seasick, including men who had previously experienced these conditions became violently ill.

It was nearly sunset when wind and wave finally subsided and weather conditions improved slightly across waters surface. (This was just wishful thinking.) Captain Barnes turned the Cat into the wind, raised the landing gear to reduce the drag and raised the wing pontoons.

Then threw the throttles forward. Waves began hitting the hull like sledge hammers. Sea water poured over the hull, blinding both pilot and co-pilot. Although raised, wing pontoons could not keep wing tips clear of waves, water poured over them.

The faithful Pratt and Whitney's roared to full power, the crew chief put the engines on "auto rich" but the sea continued clinging to the Cat. It could not attain the necessary speed to break free of the sea (80 mph). Attaining speed necessary to lift the hull up onto the 'step' could not be achieved, forces coming from continuous eight foot waves hammered the Cat into submission.

Maybe a "take off" might be possible if the Cat were to be headed downwind? Captain Barnes took time to rethink conditions. Generally speaking a pilot does not attempt a "take off" down wind but maybe it might be successful under these conditions. He turned, tail to the wind. He again pushed throttles to the firewall, rising engine RPM's produced full power. (With winds howling, engine song could not be distinguished)

Waves pushed from behind, airspeed increased. This time it worked wonderfully. At 80 mph, the Cat hit the tops of five or six swells, bounced up onto the step, hopped across a couple more swells or waves then literally jumped into the air. The final miracle: the Cat was relatively undamaged. There were no broken windows, two dozen or so popped rivets, all easily replaced, but no structural damage was found. A rugged aircraft, that Catalina.

The ride back to base took about two hours. All of us were exhausted. There were 6 crewmen and four evacuees aboard. Somehow we managed to heat a chocolate drink. Sleep came as soon as the men became horizontal.

Most evacuees were talkative. It helps to exchange thoughts with others who have experienced the same stress. Harold Steuben, the man I went after in the water, was lying in a bunk next to me. A very emotional time as our survivors had cheated death, they were still alive death had not taken them, we started talking to one another. Some people have the urge to discuss deep personal conditions. Harold was from Austin Texas, 25 years old, married and had two children.

Harold talked continuously about his family and how he wanted to return home. His time overseas had been long, beginning in 1942 with his enlistment. The man had experienced little in combat but he had navigated B24's across much of the South Pacific. He had been in constant danger and was intimately familiar with Army bases and most of the areas in the Philippine and Marianas Islands.

A B17 was waiting on the Cagdianao airstrip for our return. Survivors were quickly unloaded from the Cat and put aboard the B17 for a rapid flight to Clark Field and the Hospital. On board the B17, evacuees were given a hot meal, stripped of wet clothing, wrapped in clean, dry blankets and offered a cot for the return trip to Clark Field. Fortunately there were no major injuries, only a few bumps and bruises received during the rough loading from the turbulent sea into the Cats blister.

All crewmen who had been on the Cat with Captain Barnes were offered a hot meal upon arrival. It felt so good to know one had been part of saving lives, sleep came immediately. My bed never looked so inviting, sleep came immediately.

June 30, 1946
A Murdered Filipino

We were returning to base late one day, flying south over the Lingayan Gulf. Several miles from shore a collection of floating debris was sighted from the forward blister. We were required to investigate anything where survivors might be found from any sort of incident. Lieutenant Wells was cruising at about 8,000 feet on a southerly course, heading towards home base, Clark Field. It is difficult to identify small objects from that altitude but it was thought someone could be seen clinging to a log. Lieutenant Wells placed the Cat in a steep dive, taking the aircraft to a very close position over the debris where we could identify details. He pulled out of the dive passing only 15 feet or so above the debris. Yes, it appeared a Filipino man was clinging to a floating tree surrounded, with leaves and smaller brush. Blood could be seen on his shirt.

Lieutenant Wells circled the debris working out in his mind if or how he might attempt rescue. After circling several times he decided to make a landing. It appeared safe, meaning the survivor did not seem to be an armed Huk or Jap trying to trick us within range so he could shoot at us, explode a bomb or throw an explosive as we approached.

Ocean swells were estimated to be forming troughs about four or five feet to the crest, white caps sprayed from wave crests. This is an O K environment for a Cat to make a safe landing. Experienced Lieutenant Wells, set down almost effortlessly.

He taxied to a position beside the debris. The survivor, clinging to the log, was a Filipino who had encountered his countrymen, the Huks. A crewman jumped from the Cat to the floating log, took the Filipino in his arms and handed him to others aboard the Cat. The man was badly injured with bullet holes in his torso, burns covered much of his body, a broken wrist, there was considerable bleeding from the Filipino' scalp.

Most Filipinos living in remote areas spoke only Tagalog and little, if any English. One of the G I s, who had learned a few phrases of Tagalog, managed to question our wounded survivor who said Huks had attacked his small fishing boat in the northern Lingayan Gulf earlier in the day, shooting him, causing soon to be fatal wounds.

After the shooting, Huks threw the man into the sea and burned his boat. The Filipino managed to cling to a floating log where The Snafu Snatcher sighted him later in the day.

Since we had been on a photographic mission over Northern Luzon and not on a Snafu, we did not have a Medical Technician aboard. In need of medical procedures we were not able to give him, the man died before getting him to the military hospital. Every effort was made to save him, but there simply was not the expertise, medicine and equipment aboard our aircraft at that time to effectively treat serious injuries. It was never known where he came from, what village he called home or anything about his family. His remains were placed in an unmarked grave outside the Clark Field Military base.

July 6, 1946
The True Measure of a Commander

This Snafu was destined to become a most embarrassing incident. In addition to searching for travelers in trouble, the 2nd Air Sea Rescue was assigned the relatively easy task of taking monthly photographs of abandoned airstrips, located on dozens of islands in the Philippines. These had been constructed by both U.S. and Jap Army's. Their original purpose: to provide distressed aircraft a place to land without crashing their aircraft. Since Communist China was financing Huk Filipinos, with the intent to hijack the newly formed Democratic Government in Manila. These abandoned airstrips could be developed and used by Communist forces as a base to attack Filipino government troops.

When taking photographs, aircraft flew down one side of the landing strip. A crewman filming with a 16mm motion picture camera made a continuous strip of film of the airstrips entire length. Upon reaching the end of the run, aircraft made a sharp 180 degree turn to again pass over the airstrip while filming the other side of the landing strip. If there had been any development of military installations, it became obvious by comparing pictures taken during successive runs. It was not necessary for the pilot to be a daisy cutter that is flying within a few feet of the ground when getting these pictures, two hundred feet was low enough.

One of our pilots, I won't give you his name, made a misjudgment while filming one of these abandon airstrips. This Officer liked to hug the deck that is flying just as low as he possibly could. Would you believe only 10 feet or so from the ground while flying at cruising speed or more?

He made a misjudgment when he arrived at the end of the airstrip after making a filming pass the entire length of an abandoned runway. As the pilot came to the end of the abandon runway and was about ready to make a 180 degree turn, would you believe a coconut tree jumped into the direct path of the Cat? The port wing clipped that very tall coconut tree, the impact, ripped it off, spinning the Cat around into a crash,

Fortunately no one was hurt but the Cat was. The Cat never flew again. Damage to the airframe was so extensive no attempt was made too salvage it. However, wheels with tires, radios, radar, survival equipment and electronic equipment and firearms were all recovered. The Cats fuselage was left in the jungle, probably later stripped of all metal by native Filipinos. It was a most embarrassing moment for the pilot.

I have always wondered "what did "Jungle Jim" say to this unfortunate Officer. The story coming out of the orderly room went something like this. This pilot had acquired 80 points toward going home. A system of awarding points for combat experience had been developed to send air crews home for discharge after the wars end.

The requirement at the time 85 points were needed for the trip Stateside. "Jungle Jim" looked at the pilot's record. It was unblemished, he had flown dozens of successful Snafu's and had exhibited considerable personal bravery and skill in affecting many of those Snafu's. Surely, an Officer of this caliber should not be sent home in disgrace for an error in miss-judgment.

After careful examination of the pilots' record and a personal interview, "Jungle Jim" felt this man could not be dishonored because of a single miss-judgment. Of course "Jungle Jim" needed to "chew the man out", (that is what Officers do) which he did with the usual posturing and verbal threats of punishment. Jim did this with tongue in cheek.

He finally said to the pilot, "you have completed your punishment by simply listening to this. This stress was the same to you as five Snafus. Enduring this interview earned you the five points for needed for discharge". With a stroke of a pen, "Jungle Jim"" awarded the errant pilot the additional five points necessary for an immediate flight Stateside.

The next morning the pilot was on his way, special delivery to Manila and the replacement center. It was also rumored Major Jarnigan arranged for this man to have a seat on a Skymaster for a direct fight to San Francisco. He was to receive his Honorable Discharge with no record of his misjudgment placed in his personnel file. The pilot said, the tree jumped in front of his aircraft, he couldn't help hitting it. The Major accepted that explanation.

JULY 12, 1946
EXPLORING A REMOTE BEACH

A C47 had crash landed on the China Sea off the north coast of Palowan. Our searching efforts took The Snafu Snatchers along the north shore of Palowan, flying low barely out of reach of the splashing surf. Sometimes when there is a crash at sea, wreckage will wash up along shorelines near a crash site. This can be a productive place to search, as either survivors or debris might be sighted.

The crash site was believed to be only a few miles from the north coast of Palowan. After several days of fruitless searching the crash area, no evidence of crew or wreckage had been found. We then expanded the search area to include the northern coast of Palowan. After a day or two of searching Palowan beaches, it was decided to further expand the search to include several islands lying to the northeast of Palowan. Knowledgeable people who understood something about ocean currents in this area were always consulted before Snafu's to determine where wreckage and/or rafts, if there were any afloat, might have been carried. However, remember in those days both knowledge and tools to determine where survivors were were very limited.

When searching shorelines, slow Cats were used, flying only a few feet above splashing surf. We were looking for anything, no matter how small. A life jacket, a shred of clothing, wood, trash or anything coming from a downed airplane might be a clue we needed to find survivors.

Small, seemingly insignificant items might give some insight into a crash. Cats could fly so close, splashing surf thrown upwards, often hit the hull of a Cat. This search, starting at daybreak, had been conducted the past several hours along dozens of miles without finding any clue.

At noon, feeling the need to put feet on solid ground, Lieutenant Gerry decided to land on a small abandoned airstrip used by the military for fighter aircraft or emergency landing strips during the war. This was intended to be a break for an hour to play on the beach and/or eat a meal. This was one of many such abandon airstrips.

All materials remaining after military use had previously been either removed by the U.S. military or stolen by Filipinos. Nothing remained. This was one of the abandoned airfields photographed every month to be sure there was no renewed military activity by the Huks.

This landing strip was on a beach along a reef jutting from the coast of Mindoro. It was very simple construction. Sand had been leveled with metal interlocking sheets of steel placed across the surface of the sand. A simple but effective landing strip for smaller aircraft, intended for use when making emergency landings. Since it was constructed with little or no solid base and was short, it was intended for fighter aircraft only. Heavy bombers using it would surely over run the airstrip, ending the flight in a tangle of jungle trees.

Lieutenant Gerry decided to land the Cat and allow the crew off the aircraft for an hour of relaxation, even though landing for recreational purposes was against regulations. This required complete radio silence while we were on the ground. Since all flights were required to radio their position every hour our stop without communications with Operations on Clark Field could not last more than one hour. When communications were working properly, the location of aircraft aloft was determined by radio triangulation or radar. Remaining in one place over an hour, would have told base exactly where we were and for how long.

As the Cat approached the landing strip we passed over several warships lying on the bottom of the sea, near the shore. The ocean waters were clear and shallow, making visual sighting easy. There must have been three or four visible from the descending Cat. I learned later in life these were Jap warships.

Actually there were 24 Jap vessels of various sizes and purposes sunk by the U.S. Navy under admiral Bull Halsey's command on Sept 24,

1944. Three U.S. Submarines trapped these ships, lying at anchor, in a natural harbor. The sub's just sat underwater leisurely sinking all 24 Jap warships, one torpedo, one ship. They all went down without a fight. Some of the ships lying on the bottom had scarcely any damage to the superstructure. They had all been hit with one torpedo that punctured the hull, only once.

These several sunken ships were to become a favorite place for Scuba Divers to explore. Years later, the Philippine government set this area aside, designating it a National Park. Today a large resort catering to scuba divers exists near by. I understand regulations are strictly enforced with numerous park rangers present. Diving and exploring is permitted, however no one may enter a ship or remove any artifacts.

I busied myself hiking along the pristine sandy beach, looking at shells, starfish, dead fish, crabs and egg shells and noticing how water had shaped odd pieces of wood. Tropical beaches are rich with sea shells, mostly very different from anything else in the world. Every South Pacific Island is different. Sometimes the differences are great and obvious to a casual observer. Others are detailed and require careful and inquisitive investigation. I had entered a wonderland of new experiences and wished for more days to explore, search and wonder.

My maternal grandmother, Ethel LaValley taught me as a child to observe endless beauty in nature. She was sitting on my shoulder as we walked along the beach of Mindoro. You, the reader, might scoff when I tell you this, but actually we _were_ talking together, discussing finds, as we made new discoveries lying in the sand. Sea life in the tropics was so varied, new discoveries seemed endless. For a few moments we shared an environment, new to both of us, wondering about the miracle of the varied life we found.

There were starfish and sea urchins, both large and small. A number of blue crabs scurried from sand to water on our (remember Monnie was sitting on my shoulder) approach. One starfish was so small it could lie on a single finger. Others were larger, about a foot in diameter. As I picked up small starfish, memories returned to me. Grandmother LaValley brought a large starfish from Florida when I was a child, it was a large one, and to this day it rests in a display cabinet in my house. Sea shells came in a wide variety of shapes, sizes and colors. It would take several lifetimes to learn the intimate detail of the limitless biological variations found on the shores of tropical seas. This was certainly a new experience. The opportunity to explore some natural wonders of this earth was right there in front of me. A GIFT.

There were so many birds, it seemed hundreds of them. Since previous to entering military service I had scarcely left Odessa, New York, my knowledge of the wide variety of birds on this planet was extremely limited. Here one could see dozens of different species. Two species of gulls, a black capped tern, and shore birds of several kinds were picking their way along the beach. Pelicans were diving for fish outside the crashing surf rolling in from The China Sea. Numerous frigate birds sailed overhead.

While my understanding was limited, such experiences shaped my attitudes towards the natural world and influenced the decisions and direction taken later in my life. Yes, it was the tropics, the temperature most uncomfortable, the climate almost unbearable, but that was all forgotten in the experience of finding life, previously unknown to me.

Others of the crew had been playing, swimming and having a K ration lunch on the beach. When it came time to leave we climbed to the side hatch and returned to our respective stations aboard the Cat. We were ready to resume searching. We waited for the cough of the first engine to start.

Power to start a Cats number one engine comes from an on board battery. Power to start number two engine comes from the generators activated by the running number one engine.

The No 1 engine turned over again and again but did not fire! Oh NO! The Number one engine wasn't starting. The battery turned the engine over briskly but the cylinders would not fire. After several attempts fear began to seize us.

The battery might easily become fully discharged before the engine started, leaving us on the ground with a disabled aircraft. It would have resulted in disciplinary action if Lieutenant Gerry had found it necessary to call for assistance. This landing was unauthorized and against regulations. "Jungle Jim" would not look kindly on this violation of Regs.

We could imagine every Huk or Jap on the island coming out of the jungle, heading for us with guns blazing and machetes flashing in the sunlight. The engine finally sputtered, fired and began its song before the battery gave out.

Questions immediately asked included "was maintenance sloppy"??? The engines failure to start properly immediately put the Crew Chief in an awkward situation. Crew Chiefs are held responsible for the perfect maintenance of his assigned aircraft.

Number two engine fired the first time it turned over. Soon we were airborne, heading for Clark Field, our assigned search area for the day having been completed. Once in the air, where radar could verify our position Lieutenant Gerry made his routine check with Operations on Clark Field. "That's nice. Good boy." We failed to find any evidence of a crashed aircraft or survivors that day but Grey's eyes had been opened to a new world.

July 19, 1946
A Pregnant Patient

I was about 03:00 when the loud speaker (bitch box) called the standby air crew for an immediate flight. When they reported to Operations, they were told the Snafu was in Samar to pick up a female accident victim. The B17 on "standby" was gassed up, pre flighted and ready for immediate "take off". It was an hour later, 04:00, when the aircraft was given clearance from the control tower. Grey was seated, with seat belt firmly fastened. The pilot pushed the throttles, we started to roll.

Samar was about a four hour Flight from Clark with a small base used by the 2nd Air Sea Rescue Squadron for unexpected landings or perhaps as a fuel stop, The Air Force maintained a small base there. At times we stationed aircraft there temporarily.

This was indeed a Snafu of mercy, not to begin a search but to immediately bring an accident victim to the hospital for emergency surgery. We arrived to find the patient was a young Filipino woman who was pregnant, possibly around seven months. She also had fractures in both the right tibia and right humorous. These were injuries received in an automobile accident. The story told, she was a passenger in a Military Jeep when it rolled over. The lady was thrown from the rolling vehicle, landing several feet from the wreck. She hit the ground hard, in addition to broken bones, she suffered several severe lacerations, cuts and bruises.

One might ask why was the U.S. Army Air Force transporting a Filipino civilian from a location hundreds of miles away to the Clark Field hospital. Snafu Snatchers never learned exactly what this patients relation to the Air Force was but what little was learned, she was the wife of someone the Air Force depended on, possibly a dependable informant or someone who was somehow able to help in the constant battle with Huks who were causing disturbances on several islands.

The lady was carefully loaded into the B17. Her litter placed on the floor of the main cabin. Since she was 'shocky' her body was wrapped in blankets, her feet elevated and given hot drinks.

Our take off from Clark had been so hurried a Medical Technician was not on board. I guess one had not been included on the standby list that day. What did the rest of us know about injured pregnant ladies?

To access the extent of the ladies injuries she was stripped of her blanket and clothing exposing every inch of her skin so could be examined. Her pregnancy was obvious but what else had happened to her in that accident? While none of us were doctors, Snafu Snatchers had considerable experience treating injuries. Forget modesty, she must be given any treatment available, her life might be at stake.

Since this poor woman was in considerable pain, she was given a shot of morphine. None of us questioned the wisdom of using morphine on a pregnant woman, our ignorance could have killed her or her fetus.

Her leg had a compound fracture in the lower left femur, with visible bone protruding through broken skin. This is an extremely dangerous condition in the tropics for infections will occur in almost every untreated skin break. Additional clothing was stripped from her lower body exposing the wound. All skin breaks were cleansed and the broken femur was tightly wrapped in splints to stabilize the leg and control bleeding.

The arm was a simple fracture not breaking the skin. It was however badly "bent" giving her a source of considerable unease and pain. The arm was also wrapped in splints for protection during transport.

It was important to assess the full extent of her injuries so the radioman could call ahead to Operations on Clark Field describing the woman's condition so the hospital would be ready to properly care for her. Sulfa powder was applied, she was also given sulfa pills. This was the best that could be done for now.

With the preliminary examination made, Lieutenant Emerson guided the B17 to the end of the runway. When the airship was lined up on the runway he set the brakes and pushed the throttles to the firewall.

Four Cyclones revved up to full power, the brakes were released. It was 0800 hours when the B17 literally leaped into the air.

It was not long before this lady was placed on a gurney and wheeled into the surgical suite at the Military Hospital a few minutes before 1200 hours. Doctors were standing by, ready to give this patient immediate attention.

The lady was placed on a casting table for the doctor to assess her condition. He decided she was about 6-7 months pregnant with two broken limbs. An X ray revealed a nasty dangerous fracture in the thorax region of her back. In addition the lady suffered several lacerations and bruises. Doctors were very concerned for the fetus.

Doctors decided against performing a cesarean on the lady because there were no facilities in the military hospital to care for a 'preemie'. She would be stabilized and immediately transported to Hawaii to a civilian hospital where necessary facilities and care were available for a pregnant lady and possibly a preemie.

For now she was to be placed in a body cast. This was to be more pain for her. The doctor said he must have her awake and responsive while applying the body cast. This cast was necessary to protect her fractured vertebrae during the long flight to Hawaii.

The casting table had a hole about an inch or two in diameter in a position below a patient's crotch. An eight inch pin was placed in the hole and positioned against the vulva.

Her ankles were placed in soft anklets attached to straps that were rigged to the legs, they could be pulled tightly, forcing her legs to straighten out so the vertebrae could be properly positioned. Before the plaster cast was placed around the leg, the open wound in her femur was thoroughly cleaned, tissues repaired and stitched together.

Next, the body cast. Her shoulders were fastened to the table so she would not slide. Pressure was applied to stretch her entire body enough to straighten her back and hold it motionless while the cast was applied to both her body and femur.

My God, how that must have hurt! She was administered no anesthesia. Yet the lady uttered little sound only an occasional, barely audible grunt, could be heard. Perspiration poured from her body.

While under this pressure, the body cast was applied, with special attention given to her pregnant condition. No anesthesia was administered because the Doctor must be able to constantly monitor whether or not she continued to have feeling in her lower torso and legs. When damaged vertebra was moved, pinched nerves could paralyze abdominal organs or other nerves leading into her legs might also result in paralysis.

The doctor commented, if she tries to have this baby naturally it will surely further damage her vertebrae possibly pinching the spinal cord and cause permanent paralysis. Since her condition was critical it was decided she would be on her way to Hawaii the next day. This patient was carefully loaded on a Cat, flown to Nichols Field in Manila, transferred to a Skymaster and was in the air bound for Honolulu within a very few hours.

July 24, 1946
Jungle Rot Plagues Snafu Crews

As I have described, tropical temperatures were hostile to folks who had been born raised and had spent most of their lives in cooler, more temperate areas. Our bodies were not well adapted to the heat and humidity of the tropics. Temperatures on Luzon hovered around 100 degrees Fahrenheit. Day or night little variation existed. The only noticeable change occurred was during a few hours immediately following a typhoon when it cooled down to about 70 degrees for a few hours..

One would naturally think a cooling mid-afternoon rain would be welcome, but that was not so. It's true there are afternoon showers, rain seemed to fall every day in mid-afternoon. These showers were not cooling and refreshing. Tropical rain is hot water. When it lands on warmer ground, it turns into vapor rising a few feet where it hangs, clinging to the ground. The atmosphere, turned into a steam bath, is not welcomed by anyone. Afternoon rain only added to one's misery.

Because of this continual heat and humidity our bodies were constantly wet from both the humidity and our own sweat. Wet skin provided a nursery for all kinds of beasties. Most G.I.s have little natural immunity, especially to fungus. It was impossible to keep one's body clean and dry enough to defend against these tormentors.

One of our mechanics developed a case of Jungle Rot. A fungus condition that eats away a person's skin leaving raw sores that harbor countless beasties, causing massive infections. This airman developed a severe case of jungle rot. Initially he was not admitted to the hospital rather he was treated in an outpatient clinic and returned to duty.

The area of rot on his flesh grew each day becoming larger and larger. Rot first appeared along his right side under his belt. By the time Medics decided to return him to The States, rot opened large areas of his skin, spread across his back to the center line and across his stomach to the naval area. These were very ugly and painful to the extreme. Imagine trying to work or sleep with this condition. After these infections had been cleaned up, some patients remain scared for the remainder of their lives.

August 6, 1946
Late Arriving Air Crews

It was almost an every day occurrence, our Squadron was sending aircraft on Snafu's for something other than a search and rescue. If our aircraft were not making a search, the Squadron might also be used to fly VIPs to Pearl Harbor, Japan or Australia, long flights to be sure. It seemed our aircraft were flying a non emergency mission or Snafu constantly. Sometimes aircrew's, regardless of a missions or Snafu's purpose, returned long after dark, say about 2000 or 2100 hours.

In all mess halls located in tropical areas, there was an ironclad regulation. Food not consumed during a meal was to be discarded immediately after the serving line closed. Nothing could be saved. There was good reason for this regulation as food spoils very quickly in tropical conditions. Walk in coolers, kept at 38 degrees Fahrenheit could not be considered adequate for the long term storage of once cooked food. There were so many beasties (scarcely recognized) that made food a cause for disease or food poisoning. Contaminated food can hospitalize soldiers, resulting in crippling the activities of an entire unit and/or placing many soldiers in a hospital for days.

After watching much of the military function, I wondered how we won the war? The military was so bound up in procedure, regulations and the caste system of discrimination, it seemed impossible to get critical tasks accomplished when needed.

Essential, important situations are either ignored or not realized by people making rules and regulations and playing these games. Individual soldiers, seeing critical needs and acting on their own initiative could make life easier for others, accomplished critical tasks, or simply save resources without taking additional risks..

The regulation regarding the discarding of food was ignored many days. When Operations advised the Mess Officer there would be an aircrew arriving after regular serving hours, Mess Hall personnel made it a practice to save leftover food from the evening meal.

Everything was placed in the walk in cooler immediately after closing the serving line.

Perishable food could safely be stored for the few hours necessary to feed late arriving air crews. When crewmen arrived at the Squadron area, Mess Hall personnel had set up the serving line, warmed food and were ready to serve hungry, tired fliers.

After a day spent flying, Snafu crewmen came to the mess hall for the evening meal. This opportunity was also used to discuss the day's activities, make plans for the next day and/or make any corrections in procedure needed. It was also a time when the crewmen could tell their stories and "let off some steam", as they exuberantly related their experiences and observations. "Jungle Jim" knew what he was doing when he unofficially sanctioned these sessions. Often he was present himself.

Possibly the Major understood, sometimes crewman's versions of events might differ from the ones reported in writing by Flight Officers.

August 16, 1946
Typhoon

A typhoon was sweeping northward along the west coasts of Mindanao and Mindoro. According to weather forecasts, it was to hit Clark Field about 1800 hours with winds gusting 130 to 170 mph. That is enough "breeze" to flip a large airplane like a B17 or B29 over on its back, rip down our tents and destroy the flimsy swale buildings in our Squadron area. Pray it doesn't destroy the showers, we all need them, every day.

Major Jarnigan ordered a pilot be seated at the controls of all our aircraft. A pilot at the controls could note wind changes and make corrections necessary to keep the aircraft headed into the wind by using the rudder and yoke. When typhoon winds are blowing grounded aircraft will respond to controls as if it were actually flying. When a hurricane, typhoon or other high wind regardless of name passes a given location, wind directions change. Should shifting wind direction hit aircraft from either port or starboard the aircraft and get under the wing, it can easily be flipped or overturned causing extensive damage.

The Commanding Officer of the 313th Bomb Squadron over estimated the stability of his B29s and underestimated the strength and damage typhoon winds can do. He elected to place sandbags along the tops of his B29s wings in an effort to hold them down. He did not place a pilot at the controls. His reasoning: B29s were heavy enough winds could not damage them extensively by "flipping" them over..

When the typhoon hit, several B29s at the 313th Bomb Squadron were flipped over landing upside down on their backs, severely damaging them. It is indeed an ugly sight to see B29s lying on their backs, tail crushed, propellers bent, its bottom exposed and wheels spinning aimlessly in the air. In contrast, although our aircraft were smaller and lighter we suffered NO loss or damage to our aircraft. Remember, B29s, were about three times the size of our PBYs or B17s.

At 1800 hours (6pm), the typhoon was predicted to arrive within an hour with winds gusting to 130 mph. However, at 1800 hours there was no wind, skies were clear, the sun was setting.

In my youthful naive experience, it seemed difficult to understand how a typhoon could be so close and coming our way with no clouds in the sky and the air so calm? It seemed like a good time to take a nap.

At 1900 hours (7pm), the typhoon hit Clark Field, just as predicted. The sky had turned black while I napped. Winds were howling and heavy rains, driven by the wind, were inundating the Squadron area. As winds increased, it seemed our tent was doomed. It swayed in the wind appearing about to "take off" like one of our airplanes. At first the four of us huddled against the screaming wind. Soon we decided a nearby ditch might be a safer place to 'ride out' the storm. With ponchos shielding our faces, we fought our way out of the tent through driving rain and overpowering wind to our destination, a nearby ditch.

We dropped into the ditch, now filled with running water. At least it was clean, muddy water. Driving rain flushed out decaying garbage that ordinarily lay in the ditch before we dove into it. While it was a messy, wet and dirty place to be, it did provide safety. Probably the greatest danger during wind storms comes from flying debris. All sorts of objects fly. For instance, a sheet of flying corrugated tin roofing can severely cut a person enough to kill if it hits them. A flying 2x4 length of framing can be driven hard enough to penetrate a brick wall. If such debris should hit a person, it would surely result terrible wounds or immediate death. Winds of this velocity have been known to roll trucks.

Spending a night in a ditch with cold water flowing around you is not pleasant. The endless night passed while we were lying in the ditch listening to screaming, shrieking winds. Sleep? Forget it. Overhead, stuff was flying, we lie hugging the ditch, hoping none would hit any of us. Yet while lying in the ditch was very uncomfortable, it did seem safe. I noticed winds did not seem to bother within a foot or so of the ground. Actually the air seemed almost still in this small space close to the ground. Above, the wind was ripping, everything airborne was flying at typhoon speed which I understand contained gusts of 140 mph that night. Wind driven rain stung and cut exposed flesh, in defense we pulled ponchos over our heads. I tried to remember God was watching over me. Yeah sure. I could only hope HE wasn't also hiding someplace himself to dodge the fierce winds.

With first light of dawn a scene of destruction greeted us. Most tents had been torn apart. While damaged, ours fortunately was still erect, with torn sections of the pyramid section missing. It surely needed some repair work. But it was still standing. The thought went through my mind, the Filipino ladies on base would have some employment with needle and thread putting things back together again.

The Motor Pool had been a substantial frame building. However, it was completely destroyed and in shambles. The Mess Hall was OK, untouched actually. Fallen trees blocked roadways everywhere. Miscellaneous debris littered the ground including sheets of tin used for roofing. Most anything might be found on the muddy ground, things like clothing, a shoe or two. Even personal papers might be scattered in obscure places. Would you believe a box of condoms survived and was found on the ground under the space where a tent had sat. What are the little lambs doing?

It was my day to open the mess hall and have breakfast ready for all in the Squadron. There was little time for personal clean up. Hoping the mess hall was still standing I headed through debris to the mess hall with first light. It was indeed a wonder, it was still standing, with little apparent damage.

All Filipinos reported for duty as scheduled. None of them commented on their personal losses, they were here to feed the troops. One of the GI cooks was a little slow as his tent was probably over on the east side of Luzon now.

He had nothing of his personal possessions remaining, FORGET IT, we had hungry soldiers in shock to feed. Soon the stoves were hot. Spam and eggs ready to be served on the line.

The first soldiers to arrive were fed, as usual with expected gripes about SPAM again. One of my co-workers reminded gripers, Spam was better than nothing.

The Base Commander ordered all to be restored as soon as possible. Additional civilians were hired to clean up the area. Soon, Filipinos were swarming over the base cleaning up debris and removing fallen trees. Others were put to work repairing buildings and re-erecting tents. Easy and cheap to make, swale buildings were replaced within a few days.

August 22, 1946
A New Base Commander

It was about that time a new base Commander was assigned to Clark Field. It was later learned he was a recent graduate of West Point, long on 'book learning' and short on implementation with men. The war was finished, any resistance from surviving Japs not officially recognized nor were encounters with Huks acknowledged. The newly assigned Base Commander, a recent graduate fresh from West Point, thought the informality of fighting troops should be changed into the spit and polish of a regular military, commonly known among the draftee's as 'chicken shit'. This meant white glove inspections once weekly, Saturday morning, close order drill formations on the parade ground and a more formal control of passes enabling a soldier to leave the base. Regulations requiring the wearing of Class A uniforms while off base were to be enforced and enlisted personnel must give a hand salute to any Officer at any time.

When this order came to "Jungle Jim's desk, he ordered one cursory inspection of our quarters (no white glove) and one Saturday formation (close order drill) on an open area behind the tents. All available men assigned to the 2nd emergency rescue Squadron gathered on an open area behind the tents for a "formation". None of us had received any close order drill practice since basic training. No one, including "Jungle Jim", had the slightest interest in such useless nonsense. Once assembled we deliberately did not appear to be a well trained unit obeying commands. We did not keep step with one another or use the feet properly when turning or reversing direction.

The day was very hot with sun beating down on us. On our heads we only had that small rag that passed for a hat placed on our heads. The formation did not stay on the parade ground very long. It was just once around the area and "Jungle Jim" called a halt. He ordered the First Sergeant to call the men to attention and dismiss them. That was enough for him. Word had it the Major called the new base Commander and told him to stuff it.

The conversation was reported to go like this: To quote:

"My people work around the clock and risk their very lives answering each and every Snafu call promptly searching until survivors are found or officially given up as unrecoverable. They work all night to keep aircraft flying. I only require my men to report on time or as needed and do quality work. I issue class A passes to ALL my personnel to be kept in their possession at all times. Fatigues will be allowed any time and I will not prosecute any tickets given my men by MP's for wearing fatigues or failure to salute while off base."

The Base Commander, the newly minted 2nd Lieutenant, capitulated. One might wonder how could a shave tail from The Academy be given a post like Clark Field as a first assignment. Perhaps this demonstrated how short of experienced Officers the military was at this time. Most Senior Officers holding temporary wartime commissions had been returned home for discharge.

August 31, 1946
Not All Snafu's Have Happy Endings

Some Snafu's are indeed hard to take. There was a crash site in the jungles of northern Negros, a central island in the Philippines. There had been an extensive search conducted and the loss had been closed on the books as unrecoverable. Finding a crash site in the jungle is so difficult. Several weeks later one of our aircraft by chance sighted a break in a forest canopy as it passed close to the suspected crash site. Circling slowly, a glint of metal was again sighted while flashing through the jungle canopy. It had been so long since the crash there was absolutely no chance of a living survivor. A ground party was organized to slash their way through dense jungle to the crash site ready for whatever might be found. The Snafu, was to investigate and recover whatever might be possible.

We endured several days, fighting our way through jungle growth. This one was far from the sea, making a trek of several days necessary. Possibly it was eight days. Time becomes meaningless in such situations. It's a daily grind of taking turns on the point, being the first man to slash a hole in plant growth and penetrate the jungle. Exhausting work is putting it mildly. Not only are the physical exertion extreme, temperatures and the oppressive humidity continue to hover near 100. Supplies of food rations and water were placed along our trail, lightening our load yet leaving supplies available when returning. With ever lighter loads it was easier to make our way through the hell of this tropical rain forest. Additional supplies were dropped to us frequently, these were either carried or cached along the trail.

A person's personal discomfort becomes paramount in the mind. It became most urgent to wash up a bit. Sweat collected in clothing makes one's skin begin to feel like its crawling. One important area of a person's body that gets especially dirty and difficult to keep clean is the crotch.

It must be washed daily or skin rash will surely develop. It's a challenge how to accomplish this task. T. Sergeant Joyce, one of the crew who had been on previous jungle treks, had brought along a steel helmet (pis spot). His experience in jungles indicated there would be use for a wash basin.

This is used to take what soldiers call a 'whores bath'. Rain water is collected in the helmet and that yellow soap pad from a "Pro Kit,' used to wash the skin. This process, performed daily, is essential to keep the "little beasties" at bay.

I was taking my turn on 'point' slashing a pathway through jungle growth. There was an area where there were a number of rather large trees. Most of these trees were quite large perhaps two to four feet in diameter. Vines were everywhere, that in itself was not unusual, but here in this confined area vines, much larger then a mans arm, were growing around the trees in a spiral, all the way to the top. This seemed like a beautiful and graceful arrangement.

Mosquitoes yes, let's don't forget the pesky mosquitoes. With low levels of light, these little beasties constantly swarm around one's head biting unmercifully. Mosquitoes bring the possibility of malaria. We had been issued wide brimmed hats with netting that could be tucked under a shirt. In some ways those hats with netting added to discomfort since the netting, enclosing our heads restricted air movement from one's around face.

While approaching the crash site, the scene rapidly unfolded. It had been a Mustang fighter with only the pilot aboard. After finding the crashed aircraft, searchers fanned out in several directions looking for any clues. Hours were spent searching the jungle floor.

What was eventually found of the pilots remains? Very little. A couple of bones, a fragment of tibia, femur or humorous, a piece of skull, a belt buckle, part of a single boot and a weathered First Lieutenants bar were the only items found. A letter from home was picked up near the skull, possibly the pilot was reading it prior to his death. The paper was in poor condition, the contents of the letter barely readable. This letter was carefully protected and returned to his wife, who would likely treasure it always.

It appeared the pilot had managed to get out of the cockpit and move a few yards away from the burning aircraft before he expired. Someone voiced concern, hoping the pilot died of injuries quickly not living long enough to be attacked and consumed by scavengers or other jungle beasties. The dying process might have been prolonged and very painful. Surely, some ugly events had occurred.

An injured airman surviving a crash, cannot travel far through dense jungle undergrowth before dehydration and exhaustion overpower him. These are reasons why, in tropical areas, downed air crews must be rescued within a few days if their lives are to be saved, survival time is short.

We gathered all evidence we could find, bagged it for a return for burial. This was an important lesson. Airmen, killed in a crash in a tropical jungle will "disappear" within a few days. Very little, if anything will be recovered.

For our return to a pick up point our leader decided it might be easier to take a different route to an alternative pick up point along the seashore, he reasoned the route might be a little easier to travel.

He was correct in the decision, hacking our way through the jungle was somewhat easier as there was not the heavy undergrowth usually covering jungle floors. We entered a grove of very large trees that seemed to dominate. With my usual curiosity I started to inspect unusual appearing fallen tree limbs. Several were broken exposing the typical red of mahogany wood. This was a mahogany grove. These trees must have about 60 to 80 inches of rain to flourish. This grove seemed to extend for the next several hundred yards containing hundreds of fully mature trees.

September 8, 1946
An Accident on Clark Field

A C47 was taxiing from the runway on Clark Field to a parking apron. Along the runways and taxi strips, deep drainage ditches had been dug to carry off the heavy monsoon rainwater. Their purpose, to keep access to parking spaces and runway clear of rain water. The pilot landed and was traveling slowly along a taxi strip, turning into a parking area when he made a misjudgment. He turned too short, the port wheel on his landing gear dropped into a drainage ditch.

The spinning propeller from the number one engine hit the ground; the shaft shattered enabling the still spinning prop free to fly away. The first blade to hit the cabin, cut through the fuselage harmlessly, hitting behind the pilots seat slicing a five foot gash in the fuselage.

When the second blade hit the fuselage it had spun forward enough to come down through the outer fuselage skin just in front of the pilots sitting body. In an instant that blade severed the pilots left leg completely and nearly severed his right leg. Blood flew everywhere. Immediately an ambulance arrived, the pilot was extracted from the aircraft and swiftly taken to the base hospital. Did the pilot survive? I never knew.

A moment of carelessness or a minor distraction can result in tragedy.

September 21, 1946
Take Training Seriously

Millard Young was a replacement soldier assigned to the mess hall. He was very young, barely past his eighteenth birthday. Millard was one of the last civilians to be drafted during WWII. He came to us straight from basic training, not spending a few weeks in a 'make believe' Cook and Bakers School. I remember Millard as being very green, naive and provincial. He had been exposed to the Mickey Mouse films and "short arm" inspections as were the rest of us.

He should have known better than to have unprotected sex in the Philippines. He was telling all in the Mess Hall he wanted a "Gook" girl to prove he was a 'MAN". I think he was warned about this by every one of us working in the kitchen. However, he was insistent. One could wonder, was this macho attitude the result of having a first name like Millard? That name did not necessarily describe a big macho man to some people. Perhaps he was sensitive to it, whether anyone else was or not.

One morning this child came into the Mess Hall bragging about getting 'laid' while off base in Manila. He was asked, "did he use a condom did he visit the pro station or take the girl to the medics first'? To all of these questions he said NO, I didn't need any of these, precautions are for sissies.

He did not pass the next "short arm" inspection, exhibiting definite signs of gonorrhea. He was immediately relieved of all kitchen duty and sent to the hospital where he stayed for several days. Gonorrhea was becoming resistant to the antibiotics (penicillin) used at the time. The standard treatment at the time, a series of penicillin injections were given over a period of days to affect a cure. Since Doctors did not have other antibiotics available, and this infection was well advanced when discovered, Millard remained in the hospital.

After the standard treatment of penicillin shots in the butt, Millard was returned to the Squadron but not duty. He was ordered to report to the hospital in 30 days for additional blood tests. This time, the test was positive for the presence of syphilis. Since this disease is considered highly contagious, Millard could not be allowed to work with food preparation and was considered a danger to his companions.

Millard could not be successfully treated in the Philippines, antibiotics available, were not effective in the treatment of syphilis. He was immediately returned to The States where he might be given more advanced treatment or given a Medical Discharge, "Under other than honorable conditions". Had Millard visited a "Pro Station" all would have been forgiven and he would have been returned to duty or given a Medical Discharge under "Honorable conditions". Unreported and untreated sex was a NO NO. What a shame, for this innocent and naive lad was marked for life. If he were given a less than Honorable Discharge, he would never be eligible to receive VA benefits and was probably plagued for life by his medical condition, a heavy price for one encounter with a Philippine girl.

SEPTEMBER 27, 1946
SHOPPING IN MANILA

One of my friends and co-workers in the mess hall, Bill Anderson, ran errands for the Officers Club. He held a part time job doing some of the off base errands required to make the Club run efficiently. That was one way Bill used his time off from the kitchen. He came to me one day stating he was to take a weapons carrier to Manila to buy a load of whiskey. Having no assignments that day, I agreed to go along.

It was the last week of the month no one had cash in their pocket, including me. Possibly my wallet contained enough cash for lunch and maybe five dollars more. Bill, however, had a roll of cash for use to purchase Stateside whiskey.

The trip to Manila was an adventure in itself. We witnessed a Filipino man killed in a traffic accident. A jeep, driven by a another Filipino, came roaring over an arched bridge and hit this man while he was crossing the street at the foot of the bridge, throwing him a dozen feet into the air. When hitting the ground, he never moved, not as much as a twitch. Filipinos had no doctors or a hospital available where citizens could receive medical treatment. No doubt if he were not dead already he soon would be.

Further down the highway we were again held by MP's in a safe area during a fire fight near the road ahead. This time the Philippine army had a small group of Huks cornered about a mile from the highway. The shooting was constant and concentrated in a small area.

Shooting continued several hours, eventually slowing up until it completely stopped. One if the MP's commented to us, all attacking Huks had been killed on the spot. None surrendered, just like Japs. The MP also told us the Huks arms and ammunition had been smuggled into The Philippines by the Chinese. China was very active in its determination to hijack the newly formed Philippine government using Filipinos to accomplish the job for them. If this activity were to be stopped, the United States must give substantial aid to the Philippine government.

A bridge on the only existing highway from Clark Field to Manila, damaged during the war, had finally collapsed. Traffic was forced to cross a temporary pontoon bridge, one way at a time. Hastily placed across a muddy stream, held in place with cables, it was erected by U.S. Army engineers to carry heavier traffic during the drive to free Manila. The damaged bridge could support only jeeps and foot traffic. Traffic continued to increase each month since the cessation of hostilities with Japan. A further delay.

One must not plan anything requiring a specific time when traveling roads on Luzon, as delays of various lengths of time were commonplace. The maddening thing about delays, sometimes the reason never becomes apparent to the traveler. One was left wondering "what was that about?". A delay of any length means sitting in the hot sun with no relief possible. Air conditioning in military vehicles was unheard of.

Another lesson quickly learned, always carry a full canteen or two of drinking water when making the trip to Manila, also have in your possession a supply of snacks. This was another use for the K rations and candy bars.

Conditions had improved somewhat since my previous visit to Manila the Fourth of July, 1946. Some shops were open and doing business, merchandise of various kinds was offered for sale and small, barely functional restaurants had customers and were doing a brisk business. However to my advantage, it was the week before payday and there were no Navy ships tied up at the dock, no sailors, with wads of cash, roaming the streets, looking for whatever they could find, that is souvenirs, alcohol or women. Army boys were without cash, as pay day was a week away. Merchants were very willing to make any kind of sale, just to sell something.

SNAFU SNATCHERS

Chinese are talented merchants. A shop with textiles had opened and had quite a large selection of Chinese linen. Inside I found piles of linen table cloths finished with a dragon of a thousand stitches design. I picked out one that seemed outstanding to me. But, how could I pay for it since there was so little cash in my pocket? The Chinese lady shopkeeper initially asked for more than I had available. The haggling started and I wasn't doing very well.

After some time in exasperation, I used a four letter word in Tagalog (the dialect spoken on Luzon) loud enough she could hear me. She asked "You have been here a long time Joe"?

Orientals refer to all GI's as Joe. (We referred to Orientals as "gooks"). With a smile, I replied "Been here longer than you have." Immediately her demeanor changed, the ice had been shattered, I had a new friend, she became friendly, and we talked of many things mostly about the war.

This lady had barely escaped the Jap occupation alive. She related to me how the barely survivable conditions were in China, while under the Jap occupation. It took quite some time for her to describe in detail the unimaginable cruelty of the Japanese.

Japs had driven her from her home, which apparently wasn't much, but it was home. She and her family were sent to a prison compound where they performed forced labor in the fields, producing food the Japs took for themselves, leaving little for the Chinese who produced it. Widespread starvation resulted. Beatings were daily occurrences. Jap soldiers gang raped women, young or old, often killing them after the process was finished.

It was very difficult for this naive Yankee to understand the Jap culture at the time. Frankly, I developed a pure hatred for Jap's who committed such cruel and violent behavior against helpless people. It has taken a lifetime to understand this and forgive those of Japanese ancestry, particularly those my generation.

Somehow, with the freeing of all prisoners held by Japan at war's end, this Chinese lady was transported to the Philippines. Since arriving, she had located other members of her family remaining in China, she thought they were en route to Manila where the family planned a new life.

My secret hope? She was just as she appeared, a Chinese merchant, a refuge, not connected with Huks.

In the end she practically gave me the table cloth. She asked for only two or three dollars for it, a beautiful decoration in my home for many years.

There was time for one more stop before meeting with Anderson. Nearby, a Filipino man had a table set up along the street offering wood carvings. The one I selected was a beautiful intricately carved, serving tray, of a typical village scene from spectacular tropical wood. Again the merchant was eager to make a sale, the agreed price was exceedingly low. I left with another beautiful reminder of my military service in the Philippines.

OCTOBER 14, 1946
A MOONLIGHT FLIGHT

My tent mate and buddy, Jimmy De Lisle and I were able to get a flight on a B17 to Zamboanga, one of our bases on the southern tip of Mindanao, the southern most of the Philippine Islands. The Snafu was to recover the body of a soldier who had died of acute alcohol poisoning. The remains were to be returned to The States for burial.

We lucked out. That afternoon it was a beautiful flight across several islands heading in a straight line for our target landing field. The B17 was flying at about 10,000 feet. We flew across open seas, mountains, lowlands, lakes and inland bays. The sky was clear of clouds, making easy identification of ground features possible.

We were still climbing for more altitude as Manila passed below, then it was across Manila Bay. On reaching cruising altitude, 12,000 feet our B17 passed almost directly over Corrigidor. It was difficult to visualize how that island and the peninsula of Bataan could have been the scene of that bloody battle prior to the well publicized "death march". It seemed so quiet and almost pristine now. A few minutes later, we were over the mountains of northern Mindoro. The mountain landscape was followed by a stretch across seemingly endless jungle. Lake Laujan could be sighted to port. That day was a bright blue, with placid waters; a treat for the eye to feast on.

Jungles on Mindoro passed behind and were followed by several miles of sea dotted with dozens of smaller islands. Watching the endless variations in the landscape was indeed a thrill. Around the islands, we could see bays, lagoons, reefs and volcanoes, some smoking. I never learned much geology of the Philippine islands but it appears volcanoes played a vital roll in their formation. They were formed perhaps by volcanic activity occurring over a long period of time. Many elevations suggest they were of volcanic origin, having been worn down by erosion into the rolling hilly country dominating the islands today.

From the altitude we were flying Jim, and I could not identify objects on the ground so we could not tell if any of these small islands were inhabited. Possibly they were but we couldn't tell.

At the altitude of 12,000 feet the air temperature is much more tolerable as it is much cooler and the humidity is considerably lower than nearer the surface. This makes a comfortable respite from the surface heat. Jim and I took comfort in flying at this altitude, although breathing sometimes became labored because of the lack of oxygen in the atmosphere.

This was a bright clear day with no clouds, the ocean reflecting deep blue skies. White caps on waves were not visible from this altitude, however wakes formed by ships could easily be seen and one could determine whether or not a ship was on a straight course and the direction it was headed. Visibility seemed unlimited in any direction.

While flying in the forward blister on the B17, it seemed as if one were floating over the surface of the globe. Feelings of detachment or peace, never experienced while on the ground, swept over me,

When flying a search, this is the best position on the aircraft to find wreckage or survivors. I must add here sighting a survivor is a very satisfying event. Recent sightings, made from this very B17 passed through my mind. Since this was not a search flight today, Jim and I were drinking in all the beauty The Philippines offered us.

The distance from Clark Field to Zamboanga is 700 miles. After the B17 attained altitude and reached cruising speed of 180 miles per hour, it took nearly four hours to complete the flight.

After landing at Zamboanga we were told the purpose of the Snafu, to recover the body of an airman who died of acute alcohol poisoning. He was a participant in what one might described as a fraternity hazing. He and others were trying to out drink one another. He lost.

A C47 from our base in Palowan had originally been assigned the Snafu. However a malfunction in the hydraulics resulted in the complete loss of hydraulic power. The pilot did not have the powered use of controls, i.e. no flaps, flight controls or wheel brakes. The landing gear must be hand cranked down when hydraulics are lost

Without hydraulic power, handling the yoke, controlling flaps, rudder brakes or steering meant the pilot must use brute muscle force to control the aircraft. It takes the strength of both pilot and co-pilot to muscle the controls under this condition.

This landing took all the skills this experienced pilot possessed, to accomplish. As the C47 approached the landing strip, the pilot realized he had NO flaps. Slowing the aircraft for a safe landing was practically impossible. Consequently he touched down on the runway HOT. Having no brakes, he quickly rolled the entire length of the runway.

When arriving at the far end of the runway he was still moving much too fast. What to do now? Was he going to run off the end of the runway into jungle growth? That was too messy to contemplate.

He "feathered" the Number one engine and revved up the number two engine to full power, while simultaneously kicking the left rudder. This made the aircraft virtually spin around before rolling off the runway. The aircraft started rolling slowly down the taxi strip, paralleling the runway. After taxiing at idling speed the length of the taxi strip, towards the control tower, the pilot cut the engines the aircraft came to a halt. Without flaps, or hydraulic assist, the pilot could not "take off". It was necessary to keep the C47 on the ground waiting for a repair crew, not expected to arrive for several days. We were the second aircraft sent on this Snafu to get that deceased boy home to the States.

Because of the delays, the soldier had been dead for several days, in tropical heat and no refrigeration. Decomposition had started. The body had been placed in a hand made coffin (a wooden box) with ice, melting as we loaded it on the B17. Smelly water ran out of the coffin, spilling on all of us as we were trying to load it onto the aircraft.

Prior to "take off" Jim and I went to the showers and wet ourselves, clothes and all hoping to get rid of that terrible odor. It helped but did not completely eliminate all foul odors.

Jimmy and I made sure we were in the nose blister for the entire return trip. We could get some ventilation in this forward position. Inside of the cargo area, behind the bomb bay, the air was very unpleasant. I doubt anyone stayed in the cargo compartment with the coffin for the return trip.

I will never forget the beauty of our return flight. It was sunset as we lifted from Zamboanga. The sun sat on the horizon when the pilot threw the throttles to begin "take off" from Zamboanga. The four Cyclones began their unified song, a memorable quartet as the B17 rose into the sky. While barely off the ground, the pilot made a sharp bank to port, beginning the course to Clark Field. We entered twilight as darkness started creeping over the Islands from the East.

Before the invention of radio beams, radar and compasses, stars were used by sailors to find their way across featureless seas. Grey had never been one to take an interest in stars, yes I notice them but that is about all. This night, however, an awareness of the vast numbers of stars in the sky and the wonder of them gained my full and undivided attention...

From this position, on the equator, the constellations Big Dipper with the North Star are not visible in the northern sky. Those stars are below the horizon. In the southern sky about midway to the horizon, the constellation, Southern Cross was easy to find. This was a view of the stars I never experienced before, they were blinking at me each shouting, Grey, here I am, notice me. The Big Dipper and the Southern Cross have been used by navigators since the beginning of recorded history, they point the direction of north and south. Accurate courses across trackless seas can be set, using these two constellations.

Moonlight was so bright it almost seemed like daylight. Jim and I could see the detail of islands below. The sky remained cloudless, with moonlight reflected from the sea. Long white wakes, made by cruising ships, were easily spotted on the placid sea below.

From this altitude the horizon is probably 100 or 200 miles away. I have read, at waters level the horizon is about 12 miles from the observer. Now I was looking at moonlight reflecting from water possibly, 200 miles long. The sight of this seemed overwhelming to one who can appreciate the beauty found in nature.

One seldom has a chance to see the stars as bright as we did that night. It seemed every star in the Universe was not only visible but also seeking attention. This wondrous environment put Jim and me in very reflective moods. We talked about home, our wives and future plans. Jim, an airplane enthusiast, discussed in detail, his plans to build his own light aircraft. He actually did after he returned home.

In turn, I told Jim of my interest in horses and planned to breed palominos when returning home. Together my father and father-in-law owned considerable acreage near Odessa, New York, my home. This land was available to me on request.

During moments of silence the starlight sky started talking to me. Somehow a Divine message came to me, saying HE was in charge, things would be OK with me and I would be returned safely home.

An uneasy feeling persisted, not understood at the time, yet somehow I knew severe trials were still to come but I would prevail.

We were flying at 12,000 feet across the open sea passing over numerous islands. Lights, originating in Manila were plainly coming into view as we passed over the mountains in northern Mindoro.

The navigator pointed to our position on the map. At that moment Manila lay 125 miles in front of us. A few minutes later, Manila Bay could be seen passing below. Corrigedor was slightly to port of our course, aimed directly towards Clark Field.

City lights shown brightly. (Conditions were improving in Manila). Landing lights along the runway on Clark Field, an additional 60 miles past Manila were faintly visible in the darkness of rural Luzon..

I had been treated to a rare experience that day. When I experienced what to me is exceptional beauty, it seems I have almost touched the face of God.

A long hot shower a change of clothing and a silent prayer washed the dirt and smell of the day away. A gift had been granted me that day.

October 21, 1946
Death in the Moonlight

Clark Field was being hit frequently by Huks coming to steal anything they could carry away. These intrusions generally ended in a firefight with both Huks and GI's being killed. Enlisted men in the 2nd Air Sea Rescue Squadron were assigned to sentry duty, (this was an order, not a call for volunteers) on a rotating basis to prevent thefts and equipment destruction. While there was some help from the Infantry Company stationed on Clark Field, it was our primary responsibility to guard our own aircraft. Our ships carried supplies such as food, guns, ammunition, radios, and medical supplies including morphine and navigation equipment; all easily carried off. It was a distasteful and dangerous job. However, we all had to take our turn as an armed sentry.

One night my turn came.

A jeep took me to the armory where a .30 caliber carbine was signed out to me. It was instantly apparent a .30 caliber carbine and four 15 shot clips in pouches placed in a belt, plus another fully loaded clip in the carbine gave me a total of 75 rounds, Grey this could be a serious situation you are being assigned.

My post was on the flight line guarding a Cat parked in one of the parking areas along the taxi strip leading to the runway. My tour of duty that night extended from 2400 hours to 0400 hours. The

SNAFU SNATCHERS

Cat I was assigned to guard was one I had flown on several times. During some of the Snafus the issue was in doubt, a successful result questionable. If my memory is correct, this PBY was the one I went into the water from this very blister carrying a nylon rope to tie around a drowning airman. Frankly, warm feelings towards this particular aircraft existing inside me.

The best position for a sentry on this post in bright moonlight was to stand directly under the port wing next to the landing gear where one was in shadows. Moonlight that evening was bright enough to allow for easy sighting with the .30 caliber carbine issued to soldiers while performing sentry duties at Clark Field.

Around each parking area there were deep ditches dug to drain excessive water coming from monsoons away from the parking areas and airstrips. Outside this immediate area, a barbed wire fence had been erected. That fence would not keep a man out, as it was only an easily crossed three-wire structure. Probably it was originally placed in position to simply to mark the edge of military property.

It was an exceptionally bright moonlight night, easy to sight a person or detect movement. I took a position standing directly under the wing of a Cat next to the port landing wheel. Under the shadow of the wing was a good place to stand watch. One could see without being seen. If bad people come, a sentry must see them first.

Things were so quiet it seemed danger could not be lurking nearby could it? For some time, I listened to the night sounds, Luzon is noisy at night. There was nothing identifiable to me but they sounded friendly. The Southern Cross was plainly visible in the southern sky. I first noticed this constellation from the deck of the Seacat. Now I had time to look at it with inclination to appreciate it.

Under such conditions it is easy for a soldier to become careless and not be as watchful as he could be. The belt with pouches containing four fully loaded clips reminded me, Grey, no matter how much beauty you enjoy this evening, how boring it might become or how sleepy you might be, there is great danger out there. STAY ALERT.

The sound of "creaking wire", the unique sound a wire fence makes when someone is stretching it while crossing over or passing through, was a familiar sound to this farm boy. That sound and its meaning brought me to FULL ALERT, my carbine to the ready position with safety off. The creaking of stretched wire was enough, I was ready for whatever might come.

Within a few moments a male figure, with machete in hand, could be seen walking silently up a drainage ditch directly towards me. How many more behind him? I couldn't take time to guess.

This figure was not coming to socialize with me. Possibly this one might be the first of several armed insurgents and I stood in front of them, an obstacle they must overcome.

With the flash of a naked machete my reaction was instant. In such a situation there is really no decision to be made. When the threat was about 30 yards from me, the intruder, whatever his nationality, was a dead man. The .30 cal carbine with a fully loaded clip leaped to my shoulder. The carbine ceased movement when the front blade seen through the rear peep sight covered the targets chest. The carbine spoke three times; there is no memory of squeezing the trigger. Instinct or previous experience, I suppose.

The carbine jumped three times, giving a gentle push against my right shoulder each time it discharged. The target dropped, twitched once or twice and moved no more.

Some readers might ask "Was it necessary to kill him?" When one perceives your life is at stake shooting to kill is instinctive. Those shots were taken with as less thought than shooting a deer. The target, a threat, presented itself, offering a half second at a chest, the gun leaped to the shoulder. I instantly squeezed the trigger three times, the carbine discharged, the threat eliminated. Since that night, I have been very thankful for skills learned long before that night.

Without those instincts and the confidence in the weapon, Grey would have been cut into little Grey pieces, scattered in Luzon dust.

With the sound of shots being fired, lights immediately popped on. All parking areas along the taxi strip became fully lighted, Jeeps with armed MP's appeared from everywhere. Flare guns added additional light over the body lying in the drainage ditch. An MP Officer soon appeared, asking me for a verbal report of the incident.

He agreed with my decision to shoot as I did. Huks were after whatever they could steal and never hesitated to kill or maim any GI sentry in their path. Yes, the choice was clearly, either him or me. MP's checked the body. Someone commented to me there were three hits on his chest.

"A nice tight grouping" was one MP's comment. Several other soldiers on the scene joined in to compliment my accurate shooting. I took their word for it and did not approach the body.

The MP Officer also told me there was generally a group of them when they planned a raid; they rarely acted alone. In this instance, any Huks present immediately melted back into the mountains surrounding Clark Field. I was immediately hustled back to our Squadron area and never had to take a sentry position again.

Nothing further was ever said about the incident. "Jungle Jim" or no other Officer ever interviewed me about it. No mention of it appears in my personnel file or on my discharge. Possibly politicians or the Military did not want the American public to know GI s were still in danger and dying in the Philippines? The Military has procedures to keep some incidents very very quiet.

One might wonder what my emotional reactions were to killing a man from ambush. Actually, I have had greater emotional reactions to killing a deer than shooting that Huk. I understand when one is in the comfort and safety of a comfortable position, far removed from danger it is easy to argue endlessly the justification of such a shooting. To the soldier on the scene, in perceived danger and with only a split second to make a life or death decision, this subject is not open for discussion. A decision is made, immediate action is taken. Consequences be dammed.

After that incident, I became more aware both Jap POW's and Filipino's were working in the Squadron area doing the maintenance work. All, use machetes to keep grass cut. The man I shot had been such an employee, working near my tent daily, using a machete. Previous to that night, I had treated them all with respect, stopping to talk with Jap or Flip when time allowed.

I never thought of them as a threat. With this recent experience my attitude towards "Gooks" in the Philippines changed. Many seemed more threatening. It took many years to mellow these fears and attitudes.

I hasten to add there were many good loyal, honest Filipino people working in the kitchen with me, who become close friends. To name a few, Guno, the chief cook and patriarch of the Filipino family. Marina, the little girl who taught me to open a coconut. Jose, a most important lad in the kitchen, he was in love with Marina. Esther, the mother of two children. She was a widow struggling to raise her children and an accomplished seamstress. Filipino's cared for each other, Esther fortunately had plenty of family support and finally I must not forget, Carlos. He helped me to set up and operate the food exchange program that is canned mutton for fresh fruit and vegetables. Carlos was the key player, making the exchange work.

With this military experience, I developed the ability to sleep any time of day or night, yet able to sense and respond to the slightest sound outside the tent door. This violent experience changed my reactions. If a footstep were to stop outside or touch a step to enter my tent I was immediately on my feet, combat knife in my hand. No longer would sleeping soundly day or night be possible as long as I remained in the Philippines.

What is a flash back? Do they bother healthy people? Yes, I experienced one. There is a War Planes Museum on an airfield near my home. While I knew there has been a Navy Cat in the hanger for the past several years, I never had the urge to visit it.

Then came a day I <u>needed</u> to see that aircraft. My wife and I went to the museum, entered the hanger to view the Cat.

When standing by the left main landing wheel under the wing exactly where I had stood 62 years ago a flash back hit me. Suddenly, it was all there again: the moon lit night, the squeaking wire fence, the carbine, the sight picture, the sound of bullets discharging, the push of the recoil, the Southern Cross in the sky. Conversations were loud and clear, faces of MP's, the sound of the Lieutenants voice all were recognizable elements. The complete scene, to the smallest detail was there in vivid detail.

How long did the flash back last? Probably only a few seconds, maybe a few minutes I do not know, but the memory of the flash back has stayed with me, it was so vivid the event, including every detail is still within me and will never "go away" during this lifetime.

November 1, 1946
Jap Soldiers Continue Fighting

It was November, 1946, more than a year after the Jap formal surrender in Tokyo Bay, when Lieutenant Dickinson took off, responding to a Snafu call. This was to be a most unusual situation. A squad of eight G.I.'s with Filipino Army soldiers had been sent to one of the smaller islands in the group known as The Visayan Islands. Jap "hold" outs were known to be hiding there, not content to be "quiet".

General MacArthur bypassed many smaller Philippine Islands containing Japs during the conquest of the Philippines. The strength of invasion forces hit Leyte, Mindanao, Negros, Mindoro and Luzon. Jap soldiers remaining on the smaller islands were left to either starve or surrender. The Japanese culture allowed for no surrender. The Commander of each Jap unit must make a decision, whether to continue fighting or remain quiet and hidden, hoping the Japanese Army would return.

Smaller groups of Japs refusing surrender remained on some of these isolated remote islands. Mostly they kept out of sight, not engaging Filipino troops charged with the responsibility to "mop them up". Mop up meant, capture or kill them. They must cease to be a threat to Filipino civilians.

When their rations were exhausted, it became necessary for Jap soldiers to find needed food by living off the land since there was no re-supply coming from Japan. Some food could be grown wild in the jungle but most food, clothing or clean water had to be stolen from Filipino villages. Such actions made the presence of Japs known to Filipino troops.

At wars end, G.I.s were going home. The Philippine government had been given their Independence July 4, 1946. The U.S. Army no longer took responsibility to remove hold out Jap soldiers. This responsibility was shifted to the newly formed government of the Philippines. We did, continue to supply the Philippine Army with arms of all kinds and plenty of ammunition plus more than a few 'advisors'.

Filipino troops were called to seek out Japs and engage them in a firefight if they did not surrender when offered the chance. In some situations it, was evident Jap holdouts had available, a considerable amount of firearms and ammunition. In some firefights Jap soldiers, while few in number, had superior firepower and the capability to inflict heavy casualties. Whenever there was a violent encounter resulting in casualties, Snafu Snatchers, were called to evacuate the wounded.

Lieutenant Dickinson responded to the Snafu for the evacuation of three Filipino soldiers and two G.I.'s who had been wounded in a firefight. From the brief radio report Operations received from survivors, they thought the Japs numbered about 25 men they seemed to be well armed, for the fight had been brutal, with casualties on both sides.

Since my recent experience while on sentry duty spoke well of my marksmanship, "Jungle Jim" personally asked me to go on this Snafu. He was depending on my aggressiveness and skill with the .30 caliber carbine to protect the landing party.

There were no other soldiers in the Squadron who had any combat experience or who had demonstrated any willingness or ability to use a gun. I never thought of myself as a willing killer, but the situation as explained was very compelling. Possibly my skills with a firearm could save the day for both evacuees and rescuers.

"Jungle Jim" personally made sure I was issued a quality, almost new .30 caliber carbine and several clips of ammunition. The carbine so commonly used during WW11 fired semi automatic, that is, the trigger must be squeezed for each shot. It was not a machine gun. He sent me to the shooting range to satisfy myself the firearm was accurate enough for combat action.

Again I demonstrated there was no problem for me to achieve a very tight group of five shots in a one foot bull at 100 yards from the off hand position. Major Jarnigan, our "Jungle Jim", took personal interest in difficult and/or dangerous Snafu's, taking every precaution to make sure the mission was as safe as possible and successful, without casualties.

Lieutenant Dickinson took a course to the Island described in a brief radio call to one in the Visayan Islands, lying to the west of Samar. There are dozens of Islands in this area of the Philippines. The evacuees were on a nameless rock pile of an island to the southwest of The Visayan Islands group.

Grey T. Larison

As the Snafu Snatchers approached this rock pile, we noted it was covered with jungle growth. First question, exactly where are these guys? How do we find them before the Japs do? These two critical questions needed immediate answers.

Lieutenant Dickinson brought the Cat over the island, approaching low and slow, hoping Japs would not see us if they were still hiding on this island.

The survivors were ready for us. As we flew near, one of them flashed us with a signal mirror. We now had the exact place they were hiding. We circled around, exploring possibilities for evacuation of the wounded. By radio we learned there were five wounded, three critically, so badly they could not travel on their own. They had been carried to the pick up point on makeshift litters. The Officer in the Filipino Army had two squads of men available to carry the wounded to a designated place for evacuation. When we learned the Filipinos were present in some strength we all felt a little easier about the situation.

Lieutenant Dickinson consulted with other members of the crew, together they decided on a pick up point in a small hidden cove within what appeared to be an easy walk from the evacuees' present position. Food, water and medical supplies and a map were dropped.

The Cat came into the cove low and slow. Fortunately the water was almost glassy making, a smooth landing possible. Crewmen prepared to launch three rubber rafts, lashing them together side by side.

Waters edge along the pick up point was littered with rocks, some small others large enough to damage the Cat if it hit one of them. They appeared like fragments of lava coming from some ancient eruption. These rock fragments were along the shore extending into the water.

Since they were barely below waters surface and hidden, Lieutenant Dickinson decided to keep the Cat off shore about 100 yards since there was not time enough to evaluate the depth of water close to shore. He also felt it would make possible a faster take off after we were loaded. Lieutenant Dickinson considered our party might come under direct Jap fire. Should that happen, there would not be time to pick his way through a mine field of hidden rocks trying to leave in a hurry from closer to shore, best to keep the Cat in position for an immediate departure.

The best position for a gunner was in the bow of the middle raft, watching the shoreline and tree tops for anything that might appear suspicious. Nothing was noted to indicate the presence of Jap's. Three rafts touched the island on a narrow and muddy and rocky beach. It took several minutes for the Filipino soldiers to find us as rescuers remained very quiet, fearful Jap's might hear us. They hacked a path through the jungle to the shoreline for pick up. We were not easily visible to them because of heavy jungle growth.

The wounded men were in poor condition. Their wounds were numerous and seemed to have come from several different weapons. Weren't the Japs nearly naked and starving? That was hard to believe for they put up a vicious fight. Wounds were made with .25 caliber rifles, shrapnel, grenades and bayonets or machetes.

Carefully, but swiftly wounded were loaded onto the rafts. This took precious time as the beach was muddy and slippery. The wounded remained in considerable pain in spite of the morphine they had been given. At least their companions said they had been given injections of morphine, there wasn't time to discuss the point. We thought possibly with a limited amount of morphine syringes, each syringe, containing one dose, may have been given to more than one soldier, reducing the effects. Our concern, Japs would find us and start the firefight all over again. I was the only one with a firearm. All were loaded on one of the three rafts including the G.I.s who had not been wounded.

The rafts were pushed from the beach. Crewmen paddled towards the waiting Cat. It seemed an eternity, if Japs started shooting at us while we were in the rafts, we were easy targets. My position was again in the middle raft only this time my position was in the stern facing towards shore where it was possible to shoot without endangering any of my companions.

Lieutenant Dickinson parked the Cat with engines running about 100 yards from shore. The rafts were about 50 yards from shore, when a Jap stood in plain sight along the shore and started shooting at us. He stood upright on the beach, with a clear field of fire.

This is only a guess; possibly that Jap thought there would be NO rifleman on any of the rafts. He was in for a surprise.

He was fully exposed and easily visible. The carbine leaped to my shoulder and spoke twice. My sight picture looked good at the moment the carbine discharged, as if I had a solid hit in the chest, but will never know exactly where he was hit. Familiar thuds were heard as bullets struck flesh. His shooting immediately stopped, the Jap fell backwards, disappearing in brush along the shoreline.

The rafts were still 50 yards or so from the Cat when another Jap wanted to earn his ticket to heaven. I was also happy to oblige him.

He succeeded in getting off several shots, mostly hitting the Cat, his bullets punching holes in the blister canopy and wounded a friend of mine, Dick Fudge, in the shoulder. Dick screamed in pain and fell on the rim of the blister, blood spiriting from his back. There was only time for a quick glance over my shoulder to see what happened to my friend. It didn't look good, was my friend fatally wounded?

It was at least 50 yards from me to the antagonist, an easy shot from the off hand position even with the movement of the raft.

My special skill? Snap shooting from a moving conveyance. As if I were home in the back of a bouncing pick up truck poaching deer, a snap shot aimed at his chest would immediately drop him.

While most G.I.'s did not actually hate Japs prior to the war, we learned hate while in the South Pacific. First hand accounts of Jap cruelty and atrocities had their effect on many of us. For that brief instant as the carbine whipped to my shoulder, I knew the immense power of hate as it surged through me.

My Christian training disappeared; that Jap had just shot my friend! I wanted more than to just kill him! I wanted to cause him excruciating pain before death. Belly wounds generally do not kill immediately instead, they are very painful and cause a slow agonizing death from loss of blood and or infection, especially when no medical help is available.

At this range it would have been just as easy to make a head shot killing the Jap instantly. My point of aim was deliberately low, dead center, six o'clock on his belt buckle, the carbine again cracked twice. There were two satisfying thuds as bullets hit his belly. The Jap spun around, before he fell the carbine spoke again, this third bullet hit the Jap in the butt. Thuds were immediately followed by screams of pain, then complete silence. The Jap doubled over falling into the surf, screaming, twitching and kicking in shallow water while being washed with gentle waves. The water turned dark red. He lay at waters edge writhing in agony. None of his companions cared to expose themselves while they were still within range of that carbine. I stood on the raft watching the Jap body twitch and listening to his screaming of agony, hoping he would not take the easy way out and drown, I wanted him to suffer. How is that for hate?

Where was the Filipino Army during the evacuation? They were not active during this encounter. Where did they go, why didn't they stay to protect their own? Probably they had their reasons. Maybe they kept the larger Jap force hiding in the jungle? I never understood the answers to these questions.

We had picked up a total of five evacuees: two G.I.s and three Filipinos. After loading, Lieutenant Dickinson spun the Cat around and headed for open water. Within moments the Cat with its cargo of injured, was airborne on a course for Clark Field.

Dick Fudge had become a close friend of mine after we had flown on several Snafus'. His usual position on the Cat was to operate the radio keeping contact with Operations at the Squadron.

The Flight Technician took off Dick's shirt. The bullet hit from the back, passing through the scapula, traveling diagonally to exit an inch below the clavicle. The bullet missed Dick's heart by only a couple of inches. His life was in grave danger.

Blood poured out profusely and was difficult to control. The .25 caliber rifle bullet, hitting Dick, was unstable in flight. Typically those bullets wobble in flight and start to tumble when entering flesh. Tumbling bullets cause frightful wounds, as a considerable amount of flesh is damaged when they enter the human body.

Dick was given morphine, stripped of his clothing, placed in one of the bunk beds and given plenty of food and hot drinks. With wounds like this, shock and pain were the enemies that must be controlled. Dick's care immediately became my primary responsibility.

My hovering over Dick freed the Flight Technician to give his attention to the other evacuees. Dick asked for nothing more than he was being given. Upon arrival at Clark Field, Dick was one of the first to be loaded into an ambulance for transport to the hospital where he was immediately examined by triage nurses and taken to the operating room. I stayed close to watch a friend during his painful trial.

Doctors could repair damaged flesh and shattered bones but not nerves. The shoulder wound left Dick with permanent loss of full motion in his left arm and hand.

About a month later Dick was well enough for discharge from the hospital. Next he needed physical therapy, not available in the Philippines. Within a few days, Dick was flown "special delivery" to Manila, placed on a Skymaster and flown non stop to Hawaii. Eventually he was given a Medical Discharge and given a ticket to his home, Bangor, Maine.

One of the Filipino soldiers had been hit three times with rifle bullets. One hit his chest low on the right side, clipping his lung. His breathing was labored and he was blowing blood bubbles. Another bullet had hit his groin, high near the torso. This wound had apparently severed the femoral artery. Bleeding from this wound was practically unstoppable with the equipment aboard the Cat. The Fight Technician administered three units of plasma. Blood poured from his wounds as fast as it was administered in his arm. It was no use, the soldier died on route to the hospital.

The price some pay to help others. The men I knew who flew Snafu's willingly gave of themselves to save the lives of others. Everyone made a practice of going the extra mile or take any risk to preserve life. No Snafu Snatcher was ever heard complaining about the personal price paid or equipment lost to preserve life.

Emotional scars while invisible are very real, effecting ones health and attitudes for life.

November 22, 1946
Refugee Horses

Today our Snafu was to fly over Northern Luzon, taking pictures of abandoned airfields. This was only one of the methods Intelligence used to monitor these many abandoned strips. Maybe the Japs had built them or maybe we had? It was actually of no importance who had built what, but at that time, all abandoned runways had to be checked and photographed every week to be sure the Communist Chinese backed Huks were not developing an airfield for their own use.

While it was never acknowledged, its probable the Military was fearful China would take a more active role in supporting the Huks. To date, China had only made available small arms and money. However, Chinese presence in The Philippines existed as a real threat. Investigating the condition of usable runways was not a "make work" pastime.

We left Clark Field, flying north towards the mountains near Baguio, the summer capital of the Philippine government. Flying around 8,000 feet, the Cat was passing through broken clouds. Actually it was a memorable experience, as the mountains below were beautiful. Puffy white clouds drifted around higher elevations, making most evidence of humans invisible. We made passes over a couple of abandon and remote landing strips, filming without incident, noting nothing appeared to be changed. No evidence of unauthorized use existed. We flew further northwestward towards the Lingayan coast, where a runway lay beside the water near the village of Loag.

Lieutenant Sewell, the pilot took the Cat down to fly about 50 feet over the center of the runway. It appeared an additional building had been erected near the location of what had been a refuge shack. This had been a landing strip for emergencies only. Another pass was made, more pictures were taken.

No people had been noticed on the ground. Suddenly something hit the Cat with a sound that rattled all aboard, the Cat shuttered, cameras continued to roll, taking more pictures. Another bang. Something had hit us again. Since the Cat seemed undamaged and was functioning okay, we continued the mission.

What were the sounds? Gunshots? After landing at Loag, a visual inspection revealed two bullets had hit and penetrated the outer tip of the wing near the pontoon. The war wasn't over yet! Some Huk was making war with a Rescue Squadron.

Later, in his flight report Lieutenant Sewell detailed the incident. Infantry soldiers were dispatched to the airfield, sure enough there was some major Huk activity in the area. An active airfield was in operation not far from this abandon airstrip, possibly it was a target for the Huks since it too was quite remote.

Leaving Loag, the Cat passed over the north shore of Luzon, crossing twenty miles of open water across the Babuyan Channel and on to Fuga Island.

Fifteen miles to the north lay Dalupiri Island and twenty miles to the East we crossed Camiguin Island. None were very large and lie scattered across the water. While flying over Fuga Island a herd of horses could be seen running from the sound of our Cat.

We had been flying at around 1,000 feet. I crept forward, entering the flight deck to ask Lieutenant Sewell to fly low and slow over the herd to get a better view of them. He was aware of my interest in horses and was willing to accommodate me.

Lieutenant Sewell took the Cat down to less than 50 feet above the herd, flying lazy circles around the running horses. (In retrospect what I could have accomplished with a camera with that photo op).

My best guess, there were about 40 horses in this herd. These horses were all solid colored, with no paints, blankets or leopards present, mostly they were bays with black manes, sorrels and a few chestnuts. There was a small shack and a dock on north side of the Fuga Island. None of us were able to distinguish corrals or fencing.

The island did not appear large enough to sustain such a large herd. It seemed likely they were not wild, but part of a ranch operation. The horses were similar to those previously seen on the streets of Manila, except from this altitude this herd appeared healthier, better fed and in good physical condition. It appeared, in spite of remoteness of the islands, food was being brought to them on a regular basis.

Lieutenant Sewell glanced at the map, noticing there were other islands in this group. He indicated we could also fly to other islands to check them for more horses. We then flew slow and low over Camiguin and Dalupiri Islands. There were additional small groups of similar type horses on these islands. I have pondered about these horses since sighting them, that is where did they come from, who cares for them and how did they get to these remote islands?

Recently, my wife, Marie, while doing an Internet search for me about horses in that part of the world learned the horse was not indigenous to the Philippines, those present had been brought to these islands by various early explorers. There, in the web, she read a detailed account of the history of this herd.

A horse breeder on Luzon near the city of Baguio had been breeding and raising horses prior to the Jap invasion. He had started with the best of the Southeast Asian breeds. The Mongolian pony "type" horse was imported to the Philippines by the Chinese hundreds of years ago. Later, during the 18th and 19th centuries, Europeans introduced horses originally bred in northern Europe and Arabians and Barbs, developed in North Africa. European horses bred in cooler climates did not adapt well to the heat and humidity of the tropics. Presumably they struggled to survive, with most dying out sometime during the last century, leaving the Mongolian pony with some genes from European horses as the basic stock of The Philippines.

The breeder's goal or plan was to develop herd blood lines to improve the size and endurance of working horses on Luzon.

Before the Japs invaded Luzon, the breeder secretly began moving his best breeding stock to remote uninhabited islands lying several miles offshore, north of Luzon.

He sensed war was coming and feared Jap soldiers would either steal, work to death or kill his horses for food, destroying his life's work. At the time of the Jap invasion, the islands to the north were almost unknown and were uninhabited.

The breeder hoped, since these Islands had no military value, Jap military would ignore them and the Japs might never find his horses hidden on remote islands. Obviously the Japs never found this herd of horses for here they were a year after wars, end running fat and free on Fuga, Dalupiri and Camiguin Islands.

The breeder's plan for these horses, developed after the War, was to cross the 'pony' Mongolian type with those imported from Indonesia. Indonesian horses had been brought from Europe by the Dutch several hundred years ago, many of them being developed from Andalusian, Barb and Arabian blood lines, horses previously adapted to the heat of Africa's desert. The breeder was seeking to develop horses better adapted to tropical environments and strong enough to either ride or be a beast of burden. He thought there would be a strong market for domestic horses on Luzon at war's end.

Wars end brought changes in the need for horses in the Philippines. Horses were no longer used as beasts of burden as the world shifted to mechanized vehicles when they became available. When the economy of The Philippines improved, working horses were no longer used. However the horse racing industry developed rapidly after the war, through the 50's and 60's. This herd, we were watching from the air, eventually became foundation stock, producing horses bred exclusively for racing in the Philippines.

To protect their infant industry, The Philippine government restricted the importation of most other horses. Horses bred for racing are carefully monitored by the Government and cannot be imported into the Philippines.

December 10, 1946
"Jungle Jim" Says Goodbye

Several weeks passed, "Jungle Jim" was in the mess hall one day chewing on me about something or other minor. During the conversation he called me "soldier". That was a term I had come to despise. I replied, "Sir (always say Sir loud and clear when addressing any officer especially when he might not like to hear what you have to say) I am no soldier. I am a civilian in this uniform" He cut the conversation, spun on his heel and left the mess hall. His shoulders were moving or shaking, I'm guessing probably "Jungle Jim" was laughing all the way to his office.

While processing out of the Squadron weeks later, in preparation for my return home, a call came over the PA system "Larison to the orderly room". After first sputtering under my breath "what the hell do they want now" I immediately reported as ordered. On presenting myself to the First Sergeant, he grunted, The Major wants to see you. What did the Major want, Sarge refused to say? What is the problem now? Probably very little.

In the past, the Major and I encountered each other several times as he was generally in Operations when I signed on an aircraft as a volunteer. He was well aware and thankful for my participation in various rescue and recovery Snafu's. "Jungle Jim" began talking to me on a man to man basis, not as an Officer to an enlisted man.

He simply wanted to say Goodbye, thanked me for my service and wishes me luck. We reminiscence over some of the Snafu's I had taken part in, especially for protecting rescuers and survivors from Jap riflemen with my carbine. He asked me what my plans were when getting home. During our conversation he remarked, "Some of us were not cut out to be soldiers.

I mentioned my wife had recently delivered a baby boy and I had a deep interest in raising horses especially palominos. We had a nice, human, personal discussion. I doubt he made a practice of sharing life with many departing soldiers.

He again thanked me profusely for my volunteer service especially my special assignment resulting in the shooting of two Jap soldiers during a Snafu.

Major James P. Jarnigan was one of the very few Officers I met who knew how to get the job done. He was a people person and knew how to motivate men to achieve spectacular results. Most of his Snafu Snatchers would willingly accept and carry to successful completion any assignment he requested of them.

The Return Home

Several of us were returning home to The States on a single shipment order. We were loaded on a duce and a half in front of the Central Administration building for the ride to Manila. As we left Clark Field, I remember thinking "visiting the Philippines might be a wonderful experience, there is so much beauty here. I began remembering the unexplored seashore, the majesty of the mountains, the hundreds of birds, acres of orchids, the beautiful people, and the blue seas. Yes, it could be a great experience, providing there was no hate, war or killing.

I said goodbye to several fine Filipino friends, while they have been out of my life for 62 years they are not forgotten, I miss them to this day.

Again my bed was assigned in the replacement area on Nichols Field. Three thousand soldiers were waiting for the ship returning us Stateside to be ready to receive us. It wasn't a long wait, I think only a day or two. There was no time for a last tourist visit into the city. The shipping order was posted, we were trucked to the floating docks for loading. (Maybe the Air Force wanted to get rid of us)

Remember, when leaving San Francisco there was a quantity of arctic gear issued to us we dropped on the dock upon arrival in Manila? Now we were required to pick it all up and carry it back to San Francisco. Such nonsense. We lugged the equipment aboard without incident, dropping it in the hold for the return trip to San Francisco.

This time there was no nice Lieutenant waiting on the gang plank before we loaded aboard, to give us an eloquent speech about the wonderful joys and advantages of a formal enlistment in The United States Army Air Corps. Perhaps the Official attitude might be something like this. "If you don't want to be here go home! We don't want you we are finished with you".

The ship, waiting at dockside, for return to San Francisco was The General Hughes, a faster, more stable ship than the Seacat. The General was larger, had twin screws, and stabilizers.

Space for the troops was a bit more comfortable than our previous sea voyage. Not great mind you, just a trifle better. To be specific, there were four bunks on top of one another on "The General", not five as on the Seacat. Stabilizers reduced the violence of the ships roll, pitch and yaw while sailing across Open Ocean. Reducing the amount of roll, pitch and yaw of the ship reduced the occurrence and severity of seasickness among the passengers. The return trip was to be faster as with more powerful engines, twin screws and a sleeker hull the General Hughes had a faster cruising speed, about 23 mph.

Lines were dropped, the General backed into Manila Bay, our return journey started. I remained topside watching Corrigidor and Bataan pass to starboard, noting more islands from the General railings as we passed through channels leading to the open Pacific.

Easily visible from the port railing, our ship passed Bulusan Volcano spitting mostly smoke, with a small amount of fire. Immediately after passing the volcano, The General changed course slightly to port, threading between the Islands of Luzon and Samar. The course then shifted a bit to starboard and in a few moments, The General entered the Philippine Sea and on to the Pacific Ocean.

Working my way past other soldiers standing on the deck while saying goodbye to The Philippines, I arrived to the fantail. The Islands were, as usual covered with clouds, hiding them from view as they were left astern. Soon islands evaporated into the clouds and were no longer visible. Goodbye Philippines, you helped me to grow up

Each of us had received varying experiences some willing to talk about them and share their experiences, others were not. Among the passengers were Infantry, Air Corps and Marines, going home for discharge.

While the General was faster than other troop transports, we all remarked "the General is too slow". A comment I remember an Infantry man making "I can swim faster than this tub can move". It is well known, soldiers must complain about something,

Yes, I will agree, the return trip across the Pacific did seem to take forever. (Actually, the eastbound return trip was three days fewer than when we were westbound on the Seacat). Without exception, these men were filled with happy thoughts and were looking forward to a joyous reunion with their families, especially their wives or girl friends. The long days of unpleasant and/or dangerous duty and uncertainty were rapidly receding into history.

Fantasies filled my head about the things my wife, Ginny and I would be doing as soon as I was at her side. I plotted endlessly about exactly how to make the trip from the Elmira, New York, the train terminal and to her family farm in the hills above Cayuta, New York, where she had been living. These moments must be perfect for both of us and full of emotion, and were not to be shared with others.

As usual, all soldiers were to be assigned interior guard duty. (There was no request for a meat cutter for this crossing). A few days out of port my turn came. The post was located at the top of a stairway (companionway). Orders for the soldier standing guard at this post were taped to the wall. It instructed the interior guard to prevent all personnel, except ships crew, from using the stairway, either up or down.

Anyone going down bucked into the head of the chow line. It is very much a "no no" in the military for anyone to buck the chow or any other waiting line. This was always the rule, enforced by all soldiers at all times. One does not buck into the head of any line for any purpose. The order on the wall, however, was more inclusive. Rather than simply forbidding soldiers from going down the stairway, it stated, no one, other than ships crew was to use the stairway either UP or DOWN.

Standing guard at that post was fun. This post was on the deck near the starboard side railing, standing with other soldiers. I was engaged in constant conversations. It didn't seem like guard duty, more like just another place to socialize. Most were in good humor as all were going home after spending time in The Philippines.

Colored and white troops were separated during those years, it was not necessary to relate to each other on a daily basis. This only served to deepen anxiety between the races.

We did not understand or trust one another, when kept segregated, race relations would never improve. Even with my limited experience it had become obvious to me people must come to know each other as people. Forget race, color, gender size or any other name that divides us. We are more alike than different. Commonly noticed, we all bleed red blood and have similar emotions like love of self, spouse, children parents and neighbors. Man universally wonders about where we came from, the purpose in being here on earth and what happens after death.

This condition caused much misunderstanding. Many of the colored also called blacks, especially those from the South, spoke a language or dialect we whites could not understand. It was easy to conclude, blacks were making a deliberate attempt to remain secretive. This only added to suspicion, ignorance and distrust.

Most white soldiers were aware of the discrimination and lack of responsibility offered colored troops, but we had little understanding of how that treatment affected individual soldiers. Tension existed immediately when white and colored troops were suddenly and without warning put together.

All races slept in the same hold, but in different sections. We used the same toilets (head) and used the same showers. We stood in the same chow lines waiting our turn to be served food but very little if any fraternization occurred. None of us, black or white, were any more than civil to one another.

A colored soldier came up the stairway from the mess hall. I stopped him without making a fuss, explained what the order was and to not repeat the offense. In a few moments he returned wanting to go down the stairway. I stopped him, pointing out the posted regulations regarding the use of the stairway, an argument followed.

He was a big black man, as one might expect, when confronted in any way, he pulled a knife. It was one of the specialty knives made in the Philippines, called a Flip knife. Filipinos made them by the hundreds out of salvaged airplane parts. Knives were very pretty with colored Plexiglas or wood used to make handles and good quality steel for the blade.

These knives, sold on the streets of Manila, could be pulled from the pocket, with a twist of the wrist it would come out open giving the user a blade about four inches long ready for immediate use. The Philippine version of a switch blade.

Blacks frequently threaten with a knife even if only a verbal difference of opinion exists. Perhaps blacks think intimidation or threats will win a discussion they could not control otherwise.

Holding the open knife pointed at my belt level, he starting to make a jab at my middle. I had been armed with a lead loaded club, attached to my wrist, with a rawhide thong, hanging at my right side. The sight of the open knife brought an instant reaction from me. I brought the club up HARD, hitting him on the inside of the right humorous, inches above the elbow. A loud snap could be heard as the bone shattered. The knife dropped to the deck. My reaction, kick it overboard. Hot shot screamed, spun on his heel and left running from the scene, cradling the injured arm.

I called the Officer of the Guard and explained what had happened. The Officer replied; he is probably down at sickbay telling them he broke his arm falling down stairs. His story was not even original for that's exactly what he was telling the Corpsmen in sickbay.

When the medics finished with him, he was locked in the brig for the remainder of the crossing.

If he wants his knife back, he must swim to the bottom of the Mariana's Trench (30,000 feet surface to bottom. Water over the Mariana's Trench is deeper than Mt. Everest is high), his knife is down there. He will never threaten anyone again with it.

There were a number of colored soldiers on the General for that crossing. They started sputtering and threatening me for daring to strike one of them. I was very concerned, because any one of them could have crept into the hold where I was sleeping and cut me with a knife. However, the crisis passed without further incident.

Meaningful communications are difficult with these conditions present. Years passed before I could feel comfortable around most colored men.

I hope his arm bothers him every time it rains as long as he lives. When asked 'did I want to testify at his Court Martial in San Francisco'? This would have meant a delay of several days while the wheels of military justice turned. My reply, "I am on my way home, eager to see my wife and new baby son and do not care what happens to him. He has a broken arm to remember the incident".

The General had been at sea about 23 days bound for San Francisco. I was not privy to maps so had little idea where we were, but someone said Hawaii was behind us a couple of days past. The General did not come within sight of any of the Hawaiian Islands. We were about 5-700 miles from The Golden gate, cruising across a rolling sea.

Compelling evidence my overseas adventure was coming to an end came while standing along the port rail idly watching the stars in the northern sky, trying to identify known constellations.

As I searched the northern horizon from the port side of the ship, it occurred to me the Big Dipper should be found in the Milky Way. Looking more carefully, there it was, the Big Dipper, pointing to The North Star shining brightly welcoming this soldier home.

Drinking the reality of this Constellation took several hours, I was no longer sleepy. It was quite late in the evening when I started threading my way to the starboard railing. Sleeping soldiers were stretched out on the deck everywhere. Temperatures were cooler on deck, the fresh air, much preferable than subjecting oneself to the foul air, heat and noise in the hold.

I carefully picked my way (it would be best if I did not step on or disturb any of the sleeping men) to the starboard railing and started searching the southern horizon, looking for The Southern Cross. My eyes swept the southern sky several times; there was NO Southern Cross constellation visible. It was completely out of my view, far below the horizon.

We were in The Northern Hemisphere. This scene caused my heart to jump a few beats. The reality slowly came to me I was really out of the tropics, in a Temperate latitude, nearing home.

It was about midnight (2400 hours) two nights later when The General Hughes approached The Golden Gate. The bridge is constantly lighted. Yes, it was The Golden Gate, welcoming all returning travelers home. Lights on the bridge seemed especially bright that night.

Air travelers are denied this very emotional experience as this significant landmark remains far below and out of sight from descending aircraft. As The General approached, "The Gate," one could get the best view from the forward deck.

Every returning man aboard rushed forward to watch the bridge pass overhead as we glided below. Needless to say, it was an emotional moment. Our collective experiences made us thankful to be Americans.

I worked my way back to the fantail to watch the Golden Gate pass out of view astern of The General. I remember a feeling of gladness and relief swept over me. We crept across San Francisco Bay. The lights of the city were welcomed by many of us as they passed by on the starboard side of 'The General'. The night was finished on the steel deck.

With the first light of day, gangplanks were moved against the main deck. How good it felt to step off the gangplank and once again tread on American soil! My right foot did the honors. More than one of my companions actually bent down and kissed the dock.

Remember that arctic clothing we had been issued in Manila and required to carry back to the States? The Military had carefully recorded every item issued to us with the warning it would be checked upon arrival at Camp Stoneman, California and we would be charged if anything was missing. This proved to be just more Military nonsense. We were instructed to dump all of it in a bin on the dock, nothing was counted or checked, goodbye.

My 'victim' with the broken arm was led from the ship between MP's in handcuffs. Even through they could not court martial him because the governments witness, me, declined to stay in San Francisco, doubtless he was held in a cell until his arm healed so he too could be given an "Other than Honorable" or Bad Conduct Discharge.

What might a soldier returning from the tropics crave? Each of us had a private craving for something. With me it was real ice cream. As soon as possible I went to the nearest PX, a very complete store with a restaurant and ice cream bar. A rich chocolate milk shake made from whole milk was soon sitting before me. Man, it was rich, just what was wanted. However, consumed to fast, it proved to be too rich and would not stay down.

A few hours later, after my stomach settled a bit, I tried again, this time drinking a little slower. That was better, this time it filled that desire for REAL ice cream.

The stay at Camp Stoneman, a few miles from San Francisco Bay, was not very long, just a night or two. Enough time to be assigned to the rail car that would take us to our Separation Centers. For me, it was a return to Fort Dix in New Jersey, where this grand adventure began. Again the assignment was a Pullman passenger car with First Class accommodations. It was hooked onto a diesel powered engine that provided rapid passenger service from Chicago to points west, Stream liners prided themselves on speed. We were headed for Chicago with only three stops: Reno, Denver and Kansas City.

The return trip started with a short but slower mountain hop over the Sierras from Stoneman to Reno. That part of the journey went almost unnoticed as I slept in the Pullman berth.

In early morning the train stopped at Reno, Nevada, where we were allowed to get off and stretch for a half hour or so. There was a young girl at the station, sitting on a black and white pinto pony. We started a conversation that continued for several minutes. We talked about her horse. I also expressed my interest in horses.

She was an avid horsewoman and obviously cared a great deal for her mount. As we talked, I noticed she was of a very slight build and it appeared her back was twisted and one arm was withered as if she might be a polio victim. She did not dismount while we talked.

We enjoyed a delightful but short conversation, finished much too soon when the conductor cried "all aboard".

As events unfolded in my life, I have always wondered whether this young girl was later known as Wild Horse Annie, who lived in Reno and devoted much of her life to the preservation of wild Mustangs. I learned of Wild Horse Annie's devotion to wild horse's years later, while filming Mustangs in the Pryor Mountains, Montana She did fit the image presented in Hope Ryden's book concerning the cruelty imposed on wild horses across Western America. Hope described Wild Horse Annie as having slight build and a polio victim my friend fit that description.

The return trip across The United States was rapid, but uneventful. The Pullman car was again hooked to another Streamliner in Denver.

The Streamliner pulled out of Denver station at dusk, which would have been around 5 pm (1700 hours). While standing in the vestibule between cars during daylight hours one could watch telephone poles fly by like a picket fence. The Streamliner pulled into the Chicago rail yards shortly after midnight (2400 hours) the next night. The Streamliner had crossed the prairie Denver to Chicago in less than nineteen hours.

In the Chicago switching yard we hooked onto a local that was to take our Pullman to Fort Dix. It seemed to be the slowest machine on the planet, stopping at every city along the way. It seemingly took hours to climb the mountains in southern Pennsylvania near Altoona. Sometime in the middle of the night the car was stopped and disconnected from the engine.

We finally arrived at Fort Dix. Now the processing out of the Army Air Corps began. There was another complete physical exam given, countless lectures to endure. They took our "dog tags" and issued us NEW uniforms, these fit. Countless lectures explained Veterans benefits. At the time I was not in the mood to listen. All I wanted was that Honorable Discharge and a one way ticket to Elmira, New York in my hot fist.

Actually I had only worn fatigues since going to the Philippines and had not worn a Class A uniform except when on an official leave before shipping to the Philippines. The Ike jacket was new to the military since shipping overseas.

The military issued us new uniforms, badges and the ribbons we had earned. To be sure we had patches sewn in their proper place, seamstresses, hired by The Army were available without charge to sew patches on uniforms. That was doubtless a good move on the part of the Military as none of us had any interest in how that uniform looked. We were beyond caring whether we looked sharp in our 'new' uniforms. New or not, looking "sharp" or not, uniforms would come off, forever, upon our arrival home.

After days of processing OUT of the Military the "Ruptured Duck" was sewn on my new Ike jacket, they handed me that sheet of paper saying, in print, I was free to GO HOME.

Ginny did not know where I was, for there had been no communications since I left Clark Field. She knew I was coming home, but did not have any of the details concerning the return trip.

I hired a taxi at the train station in Elmira, and was driven the twenty five miles to her parent's farm where she had been living while I was overseas. She was asleep in bed when I entered the house around midnight.

Need I describe our joyful reunion and my first chance to hold our first child, Grey Richard? Ginny and I spent the next few days enjoying a second honeymoon. We were so happy to enjoy one another's company. It had been so painful to be denied this privilege.

The first morning I prepared breakfast for her. It was a sausage, egg and cheese omelet, the likes of which would have made my Filipino teachers proud.

Epilogue

Each of the men who served as a Snafu Snatcher was equipped not only with intelligence, courage and knowledge but a moral resolve: No person shall die because I failed" There is not one instance in the records kept by the 2nd Air Sea Rescue Squadron, in which any member of a downed aircrew was lost due to the failure of the rescue crew. I grew up, becoming an adult, in the presence of these men.

In retrospect, what happened to me during my time in Snafu Snatchers? What was learned and what intellectual and emotional changes did I experience?

When first entering military service I was a very provincial naive lad, a virgin in many ways, that is, completely ignorant of what the world is really like, having no experience outside my little provincial world. There were both harsh and pleasant lessons to be learned with severe conditions to be overcome. Fortunately my family instilled in me strong moral values. They also developed a sense of self worth with the confidence needed to overcome any situation and mature from it.

My family, could be counted on for strong emotional support through any crisis. Both my parents and grandparents were non-judgmental giving me the space to grow. This of course was a great advantage for me and was a critical ingredient during stressful encounters.

There was the separation from loved ones to be endured with a strong need to find an identity and purpose in the coming experience and make the necessary intellectual adjustments. People with different skin color, coming from cultures I had no idea existed, soon became my close friends. Most importantly I was to gain intellectual insight into how people, under stresses I could not imagine, physically survive brutal situations and make moral adjustments necessary.

FEAR
A strong almost overpowering emotion to be controlled, understood and overcome. I knew fear, the kind that can take over a person causing one to loose control and make very bad decisions.

GUILT
Other situations forced me to kill another human, once to save my own life, the second to save the lives of others. To continually harbor regrets can become debilitating. One must make peace with oneself to survive those experiences.

HATE
There was also the experience of hate, pure unadulterated hate towards an enemy. I deliberately shot not to kill, rather to cause intense pain giving my target little chance of successful recovery from the wound.

DEDICATION
I learned by the example of other men to dedicate oneself to the saving of life regardless of the cost. These learned attitudes, caused me to risk my own life in the service of others. A strong sense, we are all a part of some grand plan not understood, yes God exists. Evidence of this could be felt daily. Time after time survival depended on Divine intervention. Our personal role and reactions are always a part of it. In other situations, pain and discomfort must be endured. Most times these trials can strengthen people, if they will look for positives and learn from the experience.

Shortly after return home, interest in securing higher education developed within me. With funding available from the GI benefits Bill, a degree from Cornell University became the next goal, earned, four years later. Service in the Philippines matured me, opened my eyes to the endless opportunities to be experienced in this world. Cornell helped me to take advantage of them.

And it was not simply the availability of the educational benefits under the GI bill that motivated me to enter Cornell University. With my state of mind prior to military service, my aspirations would have never reached outside Odessa.

My service taught me all things are possible. Don't accept obstacles, overcome them. Four years later I had earned my Bachelors degree and began a life of following my dreams. This took me down many unmarked roads, doors seemed to open everywhere. Given me, were opportunities few people ever experience, a life never dreamed of.

ABOUT THE AUTHOR

Grey T. Larison served in the 2nd Air Sea Rescue Squadron affectionately called the Snafu Snatchers, based on Clark Field, the Philippines, where he took part in several rescue operations. These are his and other airmen's memoirs of sometimes very dangerous missions.

The book traces the emotional growth of a provincial country boy from upstate New York to an insightful adult who saw the ugly side of war and how lives are ended or drastically altered.

Mr. Larison had produced more than thirty environmental films, delivered thousands of lectures in public schools and written countless articles concerning the air war over the South Pacific with Japan.